Springer Tracts in Natural Philosophy

Volume 34

Edited by C. Truesdell

Springer Tracts in Natural Philosophy

Vol. 1 Gundersen: Linearized Analysis of One-Dimensional Magnetohydrodynamic Flows.
With 10 figures. X, 119 pages. 1964.

Vol. 2 Walter: Differential- und Integral-Ungleichungen und ihre Anwendung bei Abschätzungs- und Eindeutigkeitsproblemen
Mit 18 Abbildungen. XIV, 269 Seiten. 1964.

Vol. 3 Gaier: Konstruktive Methoden der konformen Abbildung
Mit 20 Abbildungen und 28 Tabellen. XIV, 294 Seiten. 1964.

Vol. 4 Meinardus: Approximation von Funktionen und ihre numerische Behandlung
Mit 21 Abbildungen. VIII, 180 Seiten. 1964.

Vol. 5 Coleman, Markovitz, Noll: Viscometric Flows of Non-Newtonian Fluids. Theory and Experiment
With 37 figures. XII, 130 pages. 1966.

Vol. 6 Eckhaus: Studies in Non-Linear Stability Theory
With 12 figures. VIII, 117 pages. 1965.

Vol. 7 Leimanis: The General Problem of the Motion of Coupled Rigid Bodies About a Fixed Point
With 66 figures. XVI, 337 pages. 1965.

Vol. 8 Roseau: Vibrations non linéaires et théorie de la stabilité
Avec 7 figures. XII, 254 pages. 1966.

Vol. 9 Brown: Magnetoelastic Interactions
With 13 figures. VIII, 155 pages. 1966.

Vol. 10 Bunge: Foundations of Physics
With 5 figures. XII, 312 pages. 1967.

Vol. 11 Lavrentiev: Some Improperly Posed Problems of Mathematical Physics
With 1 figure. VIII, 72 pages. 1967.

Vol. 12 Kronmüller: Nachwirkung in Ferromagnetika
Mit 92 Abbildungen. XIV, 329 Seiten. 1968.

Vol. 13 Meinardus: Approximation of Functions: Theory and Numerical Methods
With 21 figures. VIII, 198 pages. 1967.

Vol. 14 Bell: The Physics of Large Deformation of Crystalline Solids
With 166 figures. X, 253 pages. 1968.

Vol. 15 Buchholz: The Confluent Hypergeometric Function with Special Emphasis on its Applications
XVIII, 238 pages. 1969.

Vol. 16 Slepian: Mathematical Foundations of Network Analysis
XI, 195 pages. 1968.

Vol. 17 Gavalas: Nonlinear Differential Equations of Chemically Reacting Systems
With 10 figures. IX, 107 pages. 1968.

Vol. 18 Marti: Introduction to the Theory of Bases
XII, 149 pages. 1969.

Jorge Angeles

Rational Kinematics

Springer-Verlag
New York Berlin Heidelberg
London Paris Tokyo

Jorge Angeles
Department of Mechanical Engineering
McGill University
Montreal, PQ H3A 2K6
Canada

Mathematics Subject Classification (1980): 70B10

Library of Congress Cataloging-in-Publication Data
Angeles, Jorge, 1943–
Rational kinematics / Jorge Angeles ; edited by C. Truesdell.
p. cm. — (Springer tracts in natural philosophy ; 34)
"March 1988."
Bibliography: p.
Includes index.
ISBN-13:978-1-4612-8400-0
1. Kinematics. I. Truesdell, C. II. Title. III. Series:
Springer tracts in natural philosophy ; v. 34.
QA841.A54 1988
530'.113—dc19 88-20098

Printed on acid-free paper.

Camera-ready copy provided by the author.

9 8 7 6 5 4 3 2 1

ISBN-13:978-1-4612-8400-0 e-ISBN-13:978-1-4612-3916-1
DOI:10.1007/978-1-4612-3916-1

A Anne-Marie

Lasciate ogni speranza, voi ch'entrate.

Dante Alighieri, *Inferno*, III: 9

FOREWORD

When Prof. Truesdell kindly invited me to write a collaboration on kinematic chains for the *Springer Tracts in Natural Philosophy*, I had mixed feelings. On the one hand, it was an interesting and challenging task. On the other hand, literature on the subject is very rich, but widespread and, additionally, keeps growing at an impressive rate. As a consequence, it has become virtually impossible to give an accurate and comprehensive account of the subject. Moreover, a plethora of specialists in various fields of mechanics—finite elements, finite elasticity, computer graphics, design, robotics, etc.—appeared in the last two decades, who were engaged in some computational aspects of mechanics involving rigid-body kinematics to some extent. This subject, then, which seemed to interest a very limited number of researchers up until a couple of decades ago, suddenly became a very intensive research area. As usually happens, papers on applications devoid of any theoretical support started appearing in the literature.

At the time that I began to write this contribution, I was working on some research projects that required the devising of efficient computational algorithms for rigid-body kinematics and dynamics. It came as a surprise to me that basic questions such as *"What is the relationship between the vector of angular velocity of a rigid body and the time rate of change of the vector and scalar invariants of its rotation tensor?"* appeared virtually unanswered in the literature. Furthermore, when it came to rotation description, it appeared that the universal medium was Euler's angles, while Euler's parameters were mentioned only in passing. Other invariants, like the linear ones, that are used extensively in field theory, remained basically untouched in this context.

Thus, it appeared that a contribution of interest to those working with kinematic chains, as well as to researchers of other fields, should include a deep treatment of rigid-body kinematics, in the realm of invariant theory. and within the spirit of the *Springer Tracts in Natural Philosophy*. This is how this work came into being.

The aim of this book is, then. twofold: on the one hand, it aims at filling in gaps not covered in the literature, which are nevertheless essential to both understanding the underlying theory and devising efficient solutions to applications. These arise in areas of kinematic analysis and synthesis, and dynamics of finite-dimensional mechanical systems, i.e., systems that are representable by a finite number of state variables. On the other hand, the book introduces a unified treatment of rigid-body kinematics, and its connection with the theory of kinematic chains.

While Chapter 1 introduces some definitions, the main results pertaining to rigid-body kinematics are included in Chapters 2, 3, and 4. These are the relationships among the various invariants of the rotation tensor, their time derivatives, and the vectors of angular velocity and acceleration. Classical results like Euler's Theorem on the existence of an axis and an angle of rotation, Chasles' Theorem on the existence of a screw axis and its counterpart related to the instantaneous screw axis, are derived in a unified and rigorous manner. A further classical result, the Aronhold-Kennedy Theorem, is treated within the same framework.

Kinematic chains are given the largest chapter of the book, namely, Chapter 5. The chapter focuses on the general underlying theory and only outlines the basic problems of kinematic analysis and synthesis. Thus, items such as degree of freedom, and why the well-known Chebyshev-Grübler-Kutzbach formulae do not work in many instances, are given due attention. This leads to the introduction of the Jacobian matrix of a kinematic chain. Whereas this concept has become commonplace in the realm of open chains, its application to closed chains has remained virtually unexplored. An accurate classification of kinematic chains regarding their degree-of-freedom determination is introduced. According to this classification, criteria are derived to distinguish among trivial, exceptional, and paradoxical—also called *overconstrained*—chains. A method based on the Jacobian matrix of a closed chain is devised, that is applicable to the determination of the degree of freedom of all chains, including those with multiple loops. One more basic concept in the theory of kinematic chains, namely, isotropy, is given special attention.

This book is thus addressed to a wide spectrum of readers, regardless of their discipline, since it emphasizes the underlying general theoretical framework, avoiding, inasmuch as it is possible, a focus on a given area of applications. Moreover, the reader need not have a strong kinematics background, since an axiomatic approach has been attempted. Probably the only kinematics expertise that is required, in order to be at ease with concepts such as displacements, velocities, and accelerations, is one that can be obtained in an elementary mechanics course at the college level. Moreover, a college exposure to basic linear algebra is necessary, in order to easily follow sections dealing with ranges and nullspaces of linear transformations.

The reader used to a non-invariant treatment of mechanics will be disappointed by my decision not to include the derived relations in component form. Thus, he/she who expects cumbersome displayed equations in component form, and figures with small arrows representing infinitesimally small displacements, should be reminded of the inscription on the gates of Alighieri's Inferno: *Ye who enter here, leave all hope behind*[1]. The reason for adopting an invariant treatment of the subject is explained in the Introduction of Chapter 1. The same reader should not be discouraged from attempting to read this book, however, for it is my experience that, while understanding invariant relations in mechanics requires more intellectual effort than reading displayed tensor equations in component form, what is gained with the invariant formulation is highly rewarding.

For the successful completion of this manuscript I am indebted to several persons, institutions, and machines. Professor Truesdell deserves special thanks for his invitation. All my colleagues who have kindly supplied me with papers that are difficult to find, and with whom I have held very fruitful discussions, are highly acknowledged. I refrain from including names here because the list would be incomplete. The material of Chapters 2 and 3 was developed as a result of discussions I held with Messrs. A. A. Rojas and K. Spring, both former students of mine, the latter of whom tried to convert me into quaternions while producing a thoughtful and comprehensive review on the subject. Dr. Carlos S. López-Cajún, currently a Research Associate at the McGill Research Centre for Intelligent Machines (McRCIM), is given due acknowledgement for his help in proofreading the manuscript and his valuable criticism. Dr. López-Cajún did not only this, but also helped in editing the overall manuscript. Mr. C. Gosselin, a graduate student under my supervision, and Mr. C. Balafoutis, a graduate student at Concordia University, contributed with a highly critical proofreading. Ms. Lenore Delgado, who graciously volunteered to read the manuscript, is herewith given a cordial word of acknowledgement. Furthermore, the Department of Mechanical Engineering, of McGill University, and the McGill Research Centre for Intelligent Machines are to be thanked for the support that they offered me throughout this task. The funding supplied by the Faculty of Graduate Studies and Research of McGill University, the Fonds pour la formation des chercheurs et l'aide à la recherche, of Quebec, and the Natural Sciences and Engineering Research Council, of Canada, is highly acknowledged. Of course, the moral support and the inspiration provided by my wife Anne-Marie and my children Patricia, Bruno, and Romain, played a definite role in completing this manuscript. Last, but by no means least, "old" VAX 11 780 of the McRCIM computer network, the laser printer connected to it, the terminals both in my office and at

[1] Translation taken from Minchin, J. I.(1885) *The Divine Comedy of Dante Alighieri*, Longmans, Green, and Co., London

home, and the INRSTEX typsetting system are to be acknowledged for their fine work. Certainly, these are not to blame for misprints that the careful reader may still find in this manuscript, although it is my belief that the joint effort of many has contributed to reduce these to a minimum.

Outremont, Québec, March 1988.

Jorge Angeles

Contents

Chapter 1 PRELIMINARY NOTIONS

1.1 Introduction

A *rational* study of kinematics is a treatment of the subject based on *invariants*, i.e., quantities that remain essentially unchanged under a change of *observer*. An observer is understood to be a reference frame supplied with a clock (Truesdell 1966). This study will therefore include an introduction to invariants. The language of these is tensor analysis and multilinear algebra, both of which share many isomorphic relations, These subjects are treated in full detail in Ericksen (1960) and Bowen and Wang (1976), and hence will not be included here. Only a short account of notation and definitions will be presented. Moreover, definitions and basic concepts pertaining to the kinematics of rigid bodies will be also included.

Although the kinematics of rigid bodies can be regarded as a particular case of the kinematics of continua, the former deserves attention on its own merits for several reasons. One of these is that it describes *locally* the motions undergone by continua. Another reason is that a whole area of mechanics, known as *classical dynamics*, is the study of the motions undergone by particles, rigid bodies, and systems thereof.

A popular practice that has gained many adherents in the last decades, in connection with classical dynamics, consists of first deriving the dynamical equations governing the motion of a system of particles. From these, the equations governing the motion of rigid bodies are derived by regarding the rigid body as an aggregate of particles. Since the derivation of the Lagrangian dynamical equations for systems of particles is rather simple—no rotation is involved—, those of rigid bodies are usually derived by sheer summation of forces over a set of infinitely many particles. As pointed out to this author by C. Truesdell in a personal communication, Kirchhoff (1876) appears to be the first one to introduce this approach, which is by no means free of flaws. The popularity of this approach is surprising, for the founding fathers of classical dynamics, i.e., Newton, Euler, and D'Alembert (Szabó 1977), treated rigid bodies as continua, and not as aggregates of an infinite number of particles. Indeed, Euler (1765) begins with the ideas of mechanics in general. After that, he considers the motion of a single particle at some length, but states nothing about systems of particles. Part II of this reference treats rigid bodies using integrals over a continuum from the start.

Classical dynamicists of the turn of the century followed the rigorous formulation of deriving rigid-body dynamical equations from the general principles of conservation of momentum and angular momentum (Timerding 1902; Schoenflies and Grübler 1902; Henneberg 1903; Jung 1908; Stäckel 1908; Whittaker 1927). It is also recalled that Hamel despised mass points, and introduced in (Hamel 1912) a chapter on the *what is called mechanics of points*—German: *Die sogenannte Punktmechanik*—, but only after his statement of axioms for mechanics as a whole.

However, the derivation of Lagrange's equations for finite-dimensional mechanical systems composed of discrete sets of masses and rigid bodies was given only recently by Wang (1979), whereas a thorough study of rigid-body kinematics and dynamics was given by Wittenburg (1977). The latter introduced a method for deriving a set of constrained dynamical equations for the systems at hand, based on graph theory, that avoids dealing with the formalism of Lagrangian dynamics. Kane's formulation of Lagrange's equations (Kane 1968) is a notable contribution in the development of classical dynamics. Kane first introduced his method in 1961, in connection with nonholonomic systems (Kane 1961).

It appears that the reason why many a book on classical dynamics follows Kirchhoff's approach is a lack of understanding of the kinematics of rigid bodies. Thus, one finds extensive discussions on ill-defined—or, sometimes, totally undefined—esoteric quantities such as *quasi-coordinates* and *virtual displacements*. This confusion may have arisen from the handling of kinematics relations in component form, the readers, and the authors themselves, thereby getting lost in jungles of never-ending equations involving components of vectors and entries of matrix representations of tensors. These cumbersome equations obscure the actual significance of the physical and geometric relations they intend to illustrate. The need to clarify kinematic relations among rotation tensors, their invariants and their time derivatives, together with the concepts of angular velocity and acceleration, is one more motivation behind the writing of this book.

The kinematics of rigid bodies owes its establishment, to a great extent, to Euler (1758, 1765, 1776). Further work on the subject is due to Chasles, Rodrigues (1840), and Cauchy, in the nineteenth century. Study (1903), Dimentberg (1978), and Bottema and Roth (1979) are among the most outstanding contributors to the development of the subject in this century.

The kinematics of rigid bodies has attracted the attention of mathematicians and engineers. The field of applications in which the latter have profited from the subject to the largest extent is the *theory of machines and mechanisms* (TMM). This was established in the nineteenth century in Russia by Chebyshev (Ishlinskiĭ 1967; Artobolevskiĭ et al. 1948), in Germany by Reuleaux (1875), in Great Britain by Roberts, and in France by Savary (Hartenberg and Denavit 1964). Among the contributors to the TMM in the present century one can include the following list which, needless to say, is by no means complete: Hunt (1978) in Australia; Konstantinov (Konstantinov, Vrigazov, Stanchev, and Nedelchev 1980) in Bulgaria; Bricard (1927) and Myard (1931) in France; Meyer zur Capellen (1941) and Dittrich (Dittrich and Braune 1978) in the Federal Republic of Germany; Lichtencheldt (Lichtenheldt and Luck 1979), Volmer (1979) and Luck in the German Democratic Republic; Primrose and Duffy (1980) in Great Britain; Bottema (Bottema and Roth 1979) and Dijksman (1976) in the Netherlands; Chiang (1988) in the Republic of China; Pelecudi (1967), Manolescu and Duditza in Rumania; Artobolevskiĭ (1975) in the Soviet Union; Nieto (1978) in Spain; Freudenstein, Hartenberg (Hartenberg and Denavit 1964), Sandor (Erdman and Sandor 1984; Sandor and Erdman 1984) and Roth in the USA. Currently the TMM has attained a tremendous momentum, due to a large extent to the founding on September 27, 1969 in Zakopane, Poland, of the International Federation for the Theory of Machines and Mechanisms (IFToMM), which celebrates its world congress every four years, the most recent one in Seville, in September 1987.

Although the principles of the kinematics of rigid bodies were laid down in the eighteenth century by Euler, a dramatic development of the subject took place with the introduction of quaternions and screw algebra, in the second half of the nineteenth century. This allowed for a unified treatment of rotation and translation, and is responsible for most of the achievements recorded up to date both in theory and applications. More recently, Lie groups have been introduced in the study of space kinematics (Karger and Novák 1978). The approach adopted in the present work follows that introduced in Angeles (1982). It departs from the well-established practice of resorting to the algebra of quaternions. Instead, the author resorts to Cartesian vector- and tensor algebra and calculus, as well as linear algebra. The rotation of a rigid body will be shown to be fully described by three independent scalar invariants. However, it will be made apparent that these cannot be derived uniquely and directly from the rotation tensor involved, but via a set of four scalar invariants that observe a nonlinear dependence on the rotation tensor. This brings into play the fact that Cartesian tensor algebra alone is not sufficient to describe a rigid-body rotation uniquely. In fact, a Euclidean four-dimensional space is required to describe rigid-body rotations. A useful operation of Cartesian—three-dimensional—tensor algebra, the cross product, is not defined in vector spaces of a dimension other than three, which is the reason why linear algebra is introduced to complement the Cartesian tensor algebra and calculus. In this way, no quaternion or screw calculus is required while the possibility of handling the arising invariant relations is preserved. One item in which linear algebra is superior to quaternion algebra is its universality in the dimensions of the linear spaces it allows one to work with. Additionally, linear algebra allows transformations of a given vector space into another one of distinct dimension. Inverse transformations, in this case, are allowed via *generalized inverses*, an item that will prove to be highly valuable in deriving forward and inverse invariant relations.

1.2 On Notation and Basic Definitions

Euclidean vector spaces will be used throughout. A Euclidean vector space is a vector space of finite dimension n endowed with an inner product. It is represented here as $\mathcal{E}^n$. If $\mathbf{u}$ and $\mathbf{v}$ are both vectors of $\mathcal{E}^n$, then the inner product of these is represented as $\mathbf{u} \cdot \mathbf{v}$ in the context of Cartesian vector algebra, or as $\mathbf{u}^*\mathbf{v}$ in that of linear algebra, where $(\cdot)^*$ denotes the transpose conjugate of $(\cdot)$. The said inner product is, clearly, an element of the field in which $\mathcal{E}^n$ is defined. The real field will suffice for most of this study, and hence $(\cdot)^*$ will be changed to $(\cdot)^T$ frequently. The arising inner product becomes real under the said change.

The cross product of $\mathbf{u}$ and $\mathbf{v}$, in this order, is defined if, and only if, both $\mathbf{u}$ and $\mathbf{v}$ are elements of $\mathcal{E}^3$, which is then called the *Cartesian Euclidean space*. The said product is represented by $\mathbf{u} \times \mathbf{v}$.

A *linear transformation* $\mathbf{D}$ of $\mathcal{E}^3$ into itself is a *dyadic*, or a second-rank Cartesian tensor. Let $\mathbf{u} \in \mathcal{E}^3$ and $\mathbf{v}$ be its image under $\mathbf{D}$. This means that

$$\mathbf{v} = \mathbf{D}\mathbf{u}$$

If $\mathbf{D}$ is one-to-one, then $\mathbf{D}^{-1}$ exists and

$$\mathbf{u} = \mathbf{D}^{-1}\mathbf{v}$$

The *identity dyadic* or *identity Cartesian tensor* is represented by $\mathbf{1}$. It is defined as the dyadic under which, for every $\mathbf{u}$,

$$\mathbf{1}\mathbf{u} = \mathbf{u}$$

Every dyadic $\mathbf{D}$ has three linearly independent scalar invariants, namely, its first three *moments*, $\mathrm{tr}(\mathbf{D})$, $\mathrm{tr}(\mathbf{D}^2)$, and $\mathrm{tr}(\mathbf{D}^3)$, $\mathrm{tr}(\cdot)$ denoting the trace of its tensor argument. Thus, any other scalar invariant of $\mathbf{D}$, e.g. its determinant, is a linear combination of these three. Additionally, every dyadic $\mathbf{D}$ has the vector invariant $\mathbf{v} = \mathrm{vect}(\mathbf{D})$. Let v_i denote the ith component of $\mathbf{v}$ and d_{ij} the (i,j) component of $\mathbf{D}$; then, $\mathrm{vect}(\mathbf{D})$ is defined in index notation as:

$$v_i = \frac{1}{2}\epsilon_{ijk}d_{kj} \tag{1.2.1}$$

where ϵ_{ijk} denotes the *Levi-Civita* or *alternating tensor*, defined as $+1$ if its indices, which range from 1 to 3, appear in cyclic order; as -1 if they appear in anticyclic order; as 0 if one index is repeated.

From eq.(1.2.1), it is clear that $\mathrm{vect}(\mathbf{D})$ is a *linear* invariant of $\mathbf{D}$. Of course, $\mathrm{tr}(\mathbf{D})$ is also a linear invariant of $\mathbf{D}$. Thus, both $\mathrm{tr}(\mathbf{D})$ and $\mathrm{vect}(\mathbf{D})$ will henceforth be referred to as the *linear invariants* of $\mathbf{D}$.

Now let $\mathbf{D}^T$ denote the *conjugate* or *transpose* dyadic of $\mathbf{D}$. If $\mathbf{D}^T = \mathbf{D}$, then $\mathbf{D}$ is said to be *self-conjugate* or *symmetric*. If $\mathbf{D}^T = -\mathbf{D}$, then $\mathbf{D}$ is said to be *antiself-conjugate* or *skew symmetric*. The foregoing concepts can be extended to tensors other than Cartesian, namely to those isomorphic to $n \times n$ matrices.

The *Cartesian decomposition* of every dyadic $\mathbf{D}$ is the following:

$$\mathbf{D} = \mathbf{D}_s + \mathbf{D}_{ss} \tag{1.2.2}$$

where $\mathbf{D}_s$ is symmetric and $\mathbf{D}_{ss}$ is skew symmetric. Both are given by

$$\mathbf{D}_s \equiv \frac{1}{2}(\mathbf{D} + \mathbf{D}^T), \quad \mathbf{D}_{ss} \equiv \frac{1}{2}(\mathbf{D} - \mathbf{D}^T) \tag{1.2.3}$$

Clearly,

$$\mathrm{tr}(\mathbf{D}) = \mathrm{tr}(\mathbf{D}_s), \quad \mathrm{vect}(\mathbf{D}) = \mathrm{vect}(\mathbf{D}_{ss}) \tag{1.2.4}$$

Thus, the trace of a skew-symmetric dyadic and the vector of a symmetric one vanish. Moreover, the following relation holds:

$$\mathbf{D}_{ss}\mathbf{x} \equiv \mathbf{v} \times \mathbf{x} \tag{1.2.5}$$

for arbitrary $\mathbf{x}$. From eq.(1.2.5) it is clear that, given any Cartesian vector $\mathbf{u}$, the *unique* Cartesian skew-symmetric tensor $\mathbf{U}$ associated with it is defined as:

$$\mathbf{U} \equiv \frac{\partial(\mathbf{u} \times \mathbf{x})}{\partial \mathbf{x}} \tag{1.2.6}$$

for arbitrary $\mathbf{x}$. The foregoing discussion suggests the following relation between vector $\mathbf{v}$ and the skew-symmetric component of $\mathbf{D}$:

$$\mathbf{D}_{ss} = \mathbf{1} \times \mathbf{v} = \mathbf{v} \times \mathbf{1} \tag{1.2.7}$$

Moreover, one can readily show that

$$(\mathbf{1} \times \mathbf{v})^2 \equiv \mathbf{v} \otimes \mathbf{v} - (\mathbf{v} \cdot \mathbf{v})\mathbf{1} \tag{1.2.8}$$

where $\otimes$ denotes the *tensor product* of the two vectors beside it. Although the tensor product has been defined here for two Cartesian vectors, it can readily be extended to spaces other than the Cartesian one. In fact, the tensor product of an m-dimensional vector space into an n-dimensional space gives rise to an $m \times n$ tensor space. Now, if relation (1.2.8) is applied to a unit n-dimensional vector $\mathbf{e}$, one has

$$\mathbf{F} \equiv -(\mathbf{1} \times \mathbf{e})^2 = \mathbf{1} - \mathbf{e} \otimes \mathbf{e} \tag{1.2.9}$$

from which it is clear that $\mathbf{F}$ maps every n-dimensional vector $\mathbf{w}$ into a vector perpendicular to $\mathbf{e}$, inasmuch as $\mathbf{1} \times \mathbf{e}$ does. In fact, $\mathbf{Fw}$ is the normal component of $\mathbf{w}$ onto $\mathbf{e}$. However, $\mathbf{e} \times \mathbf{w}$ is perpendicular to both $\mathbf{e}$ and $\mathbf{w}$, whereas $\mathbf{Fw}$ lies in the plane of $\mathbf{e}$ and $\mathbf{w}$. From the foregoing relations it is apparent that every vector $\mathbf{w}$ can be decomposed into the following orthogonal components with respect to $\mathbf{e}$:

$$\mathbf{w} \equiv \mathbf{e} \otimes \mathbf{e} \cdot \mathbf{w} + \mathbf{Fw} \tag{1.2.10}$$

Furthermore, given the two Cartesian vectors $\mathbf{u}$ and $\mathbf{v}$, the following relations hold:

$$\mathrm{tr}(\mathbf{u} \otimes \mathbf{v}) = \mathbf{u} \cdot \mathbf{v}, \quad \mathrm{vect}(\mathbf{u} \otimes \mathbf{v}) = \mathbf{u} \times \mathbf{v} \tag{1.2.11}$$

The *Euclidean W-norm* of an n-dimensional vector $\mathbf{v}$ is defined as

$$\|\mathbf{v}\|_W \equiv \sqrt{\mathbf{v}^T \mathbf{W} \mathbf{v}} \tag{1.2.12}$$

where $\mathbf{W}$ is a given $n \times n$ positive definite tensor. Similarly, the *Euclidean W-norm* of the $n \times n$ tensor $\mathbf{D}$ is defined as:

$$\|\mathbf{D}\|_W \equiv \sqrt{\mathrm{tr}(\mathbf{D}^T \mathbf{W} \mathbf{D})} \tag{1.2.13}$$

Notice that, by setting $\mathbf{W} = \frac{1}{n}\mathbf{1}$, where $\mathbf{1}$ is the identity $n \times n$ tensor, the norm of the identity tensor is 1, whereas that of any n-dimensional vector is the root-mean-square value of its components. The foregoing Euclidean norm of a tensor stems from the following inner product for tensor spaces:

$$(\mathbf{D}, \mathbf{E}) = \mathrm{tr}(\mathbf{D}\mathbf{E}^T) \tag{1.2.14}$$

in which $\mathbf{D}$ and $\mathbf{E}$ are in $m \times n$-dimensional tensor spaces. The foregoing definition allows one to define the *angle* between two tensors, in a similar fashion to the definition of the angle between two vectors. Clearly, if the inner product between two given tensors vanishes, these are *orthogonal*. For example, if $m = n$, $\mathbf{D}$ is symmetric and $\mathbf{E}$ is skew-symmetric in eq.(1.2.14), then $\mathbf{D}$ and $\mathbf{E}$ are orthogonal.

1.3 Further Definitions

The subject of this work is the kinematics of motions undergone by rigid bodies. Hence a few definitions pertaining to these concepts are in order. A *motion* is a continuous mapping of $\mathcal{E}^3$ into itself, i.e., if $\mathbf{p}$ denotes the position vector of an arbitrary point $\mathcal{P}$ of the 3-dimensional space, then the image of $\mathbf{p}$ under a motion $\mathcal{M}$, denoted by $\mathbf{p}'$, is

$$\mathbf{p}' = \mathcal{M}(\mathbf{p}) \tag{1.3.1a}$$

The foregoing mapping can be ordered in time, in which case it is represented as

$$\mathbf{p}' = \mathcal{M}(\mathbf{p}, t) \tag{1.3.1b}$$

A motion $\mathcal{R}$ is said to be rigid if, given two *arbitrary* points of a set $\mathcal{B}$, henceforth referred to as a *body*, $\mathcal{P}_1$ and $\mathcal{P}_2$, with position vectors $\mathbf{p}_1$ and $\mathbf{p}_2$, their corresponding images under $\mathcal{R}$ being denoted by $\mathbf{p}_1'$ and $\mathbf{p}_2'$, the distance between $\mathcal{P}_1$ and $\mathcal{P}_2$ is preserved under $\mathcal{R}$, i.e.,

$$\|\mathbf{p}_2' - \mathbf{p}_1'\| = \|\mathbf{p}_2 - \mathbf{p}_1\| \tag{1.3.2}$$

It is pointed out that a body need not be rigid to undergo a rigid-body motion. In fact, any body, regardless of its nature, can be given a rigid-body motion by describing the position of its set of points from a moving observer. However, a rigid body can undergo only rigid-body motions.

A rigid-body motion, or a rigid motion for brevity, is in general a non-linear mapping. This can be readily shown as follows: let $\mathcal{R}$ be defined as

$$\mathcal{R} : \mathbf{p}' = \mathbf{p} + \mathbf{c} \tag{1.3.3}$$

where $\mathbf{c}$ is a constant vector. This type of motion will be referred to as a *pure translation*; then, if the images of the position vectors $\mathbf{p}_1$ and $\mathbf{p}_2$ of corresponding points of a body are denoted by $\mathbf{p}_1'$ and $\mathbf{p}_2'$, one has

$$\mathbf{p}_i' = \mathcal{R}(\mathbf{p}_i),\ i = 1, 2 \tag{1.3.4}$$

It will be first shown that $\mathcal{R}$, as defined by eq.(1.3.3), is rigid, and subsequently, that it is not additive. Indeed,

$$\mathbf{p}_i' = \mathbf{p}_i + \mathbf{c}$$

and hence

$$\|\mathbf{p}_2' - \mathbf{p}_1'\| = \|\mathbf{p}_2 + \mathbf{c} - \mathbf{p}_1 - \mathbf{c}\| \tag{1.3.5}$$

which clearly shows that the motion is rigid. Now, if $\mathcal{R}$, as given by eq.(1.3.3), is linear it is additive, i.e., $\mathcal{R}(\mathbf{p}_1 + \mathbf{p}_2) = \mathcal{R}(\mathbf{p}_1) + \mathcal{R}(\mathbf{p}_2)$. However,

$$\mathcal{R}(\mathbf{p}_1 + \mathbf{p}_2) = \mathbf{p}_1 + \mathbf{p}_2 + \mathbf{c}$$

On the other hand,

$$\mathcal{R}(\mathbf{p}_1) + \mathcal{R}(\mathbf{p}_2) = \mathbf{p}_1 + \mathbf{p}_2 + 2\mathbf{c}$$

thereby showing the lack of additivity of $\mathcal{R}$, for every $\mathbf{c} \neq \mathbf{0}$.

It will be shown in Chapter 2 that an important class of rigid motions is linear, namely that of *pure rotations.*

A *configuration* of a body is given by a set of points $\mathcal{B}$ of the body. Under a rigid motion, each point $\mathcal{P}$ of $\mathcal{B}$ is mapped into a *unique* point $\mathcal{P}'$. The set of points $\mathcal{P}'$, referred to as $\mathcal{B}'$, is said to constitute a distinct configuration of the body. The difference between the position vector of a point $\mathcal{P}'$ in $\mathcal{B}'$ and the corresponding $\mathcal{P}$ in $\mathcal{B}$, denoted by $\mathbf{u} = \mathbf{p}' - \mathbf{p}$, is said to be the *displacement of point* $\mathcal{P}$. The motion taking $\mathcal{B}$ into $\mathcal{B}'$ is correspondingly termed the *displacement of the body* under study.

1.4 Invariance

An *affine transformation* of the position vector $\mathbf{p}$ is defined in terms of an orthogonal Cartesian tensor $\mathbf{Q}$ and a Cartesian vector $\mathbf{c}$, neither of which is a function of $\mathbf{p}$. The said transformation maps $\mathbf{p}$ into vector $\mathbf{p}'$ as follows:

$$\mathbf{p}' = \mathbf{Q}\mathbf{p} + \mathbf{c} \tag{1.4.1}$$

Transformation (1.4.1) is also referred to as a *change of observer.* If $\mathbf{Q}$ and $\mathbf{c}$ are functions of time, then a change of time scale might also be involved in the foregoing transformation. This is why Truesdell (1966) defines an observer as a coordinate frame and a clock.

A *scalar function* of $\mathbf{p}$, $f(\mathbf{p})$, is said to be *frame invariant,* or simply *invariant*, if, under the change of frame (1.4.1),

$$f(\mathbf{p}') = f(\mathbf{p}) \tag{1.4.2}$$

A *vector function* of $\mathbf{p}$, $\mathbf{v}(\mathbf{p})$, is said to be frame invariant if, under the change of observer (1.4.1),

$$\mathbf{v}(\mathbf{p}') = \mathbf{Q}\mathbf{v}(\mathbf{p}) \tag{1.4.3}$$

Similarly, a *second-rank tensor function* of $\mathbf{p}$, $\mathbf{T}(\mathbf{p})$, is said to be frame invariant if, under (1.4.1),

$$\mathbf{T}(\mathbf{p}') = \mathbf{Q}\mathbf{T}(\mathbf{p})\mathbf{Q}^T \tag{1.4.4}$$

For example, let $f(\mathbf{p}) \equiv \|\mathbf{p}\|_2$, where $\|\cdot\|_2$ denotes the Euclidean norm of $\mathbf{p}$. Then $f(\mathbf{p})$ is not frame invariant. However, let $\mathbf{d} = \mathbf{p}_2 - \mathbf{p}_1$ i.e., $\mathbf{d}$ is the difference between two vectors $\mathbf{p}_1$ and $\mathbf{p}_2$. Thus, if $f(\mathbf{d}) = \|\mathbf{d}\|_2$, then $f(\mathbf{d})$ is frame invariant.

As another example, consider the vector function $\mathbf{v}(\mathbf{p})$ of the position vector $\mathbf{p}$, defined as $\mathbf{v}(\mathbf{p}) \equiv \mathbf{p}$. This function is not frame invariant, but, again, if $\mathbf{d}$ is defined as before, then $\mathbf{v}(\mathbf{d}) \equiv \mathbf{d}$ is frame invariant.

As an example of invariant tensor, consider the *second moment* of point $\mathcal{P}$, of position vector $\mathbf{p}$, with respect to point $\mathcal{Q}$, of position vector $\mathbf{q}$, defined as

$$\mathbf{P}^{\mathcal{Q}} \equiv \|\mathbf{p}-\mathbf{q}\|^2\mathbf{1} - (\mathbf{p}-\mathbf{q}) \otimes (\mathbf{p}-\mathbf{q}) \tag{1.4.5}$$

The foregoing tensor quantity is frame invariant.

As examples of frame-invariant scalar functions of a tensor argument, $\mathbf{T}$, one can cite the *moments* of a $n \times n$ tensor, I_k, defined as

$$I_k \equiv \text{tr}(\mathbf{T}^k), \quad k = 1, 2, \ldots$$

Moreover, only the first n foregoing moments are linearly independent, which is a consequence of Cayley-Hamilton's Theorem (Finkbeiner 1966; Halmos 1974). As a further consequence of Cayley-Hamilton's Theorem, $\det(\mathbf{T})$ is a linear combination of the first n moments of $\mathbf{T}$. Clearly, then, if $\mathbf{T}$ is proper orthogonal, it has only $n-1$ linearly independent invariants for, in this case, $\det(\mathbf{T}) \equiv 1$. The same holds if $\mathbf{T}$ is improper orthogonal, in which case $\det(\mathbf{T}) \equiv -1$.

1.5 Generalized Inverses

In dealing with rigid-body rotations in Chapter 2, it will become apparent that the sole algebra of Cartesian vectors is insufficient to handle them. Hence, mathematicians and mechanicists have introduced an algebra which, while containing that of Cartesian vectors, extends it in order to allow for the description of rotations. The entities that belong to such algebras have been called *quaternions* and *spinors*. Quaternions were introduced in the nineteenth century (Kelland and Tait 1882, Hamilton 1899), while spinors were first presented in 1929 by van der Waerden (Pauli 1958). Quaternion algebra has been by far the most applicable tool in the study of rigid-body rotations. An extension of this, allowing for the study of general rigid-body motions, was introduced by Ball (1900) and is called *screw algebra*. Both quaternion and screw algebras can be regarded as subsets of linear algebra. Hence the motion of rigid bodies will be studied here with the aid of linear algebra. A very useful concept in this context is that of generalized inverses for nonsquare full-rank matrices or $m \times n$ tensors. Before introducing this concept, some definitions are given.

Let $\mathbf{A}$, $\mathbf{x}$, and $\mathbf{b}$ be a full-rank $m \times n$ tensor, with $m > n$, and n- and m-dimensional vectors, respectively. The equation

$$\mathbf{A}\mathbf{x} = \mathbf{b} \tag{1.5.1}$$

cannot, in general, be solved for $\mathbf{x}$. This is due to the fact that $\mathbf{b}$ is a vector of a space whose dimension, m, is larger than that of the space to which $\mathbf{x}$ belongs, n. However, an approximation $\mathbf{x}_o$ to eq. (1.5.1) can be found which produces an error $\mathbf{b} - \mathbf{A}\mathbf{x}_o$ with a minimum Euclidean norm. The said approximation is then, the *least-square approximation* to the given system of equations. A straightforward minimization of the Euclidean norm of the error leads to the following value of $\mathbf{x}_o$:

$$\mathbf{x}_o = \mathbf{A}^I\mathbf{b} \tag{1.5.2}$$

In the foregoing, $\mathbf{A}^I$ represents a *generalized inverse* of $\mathbf{A}$, which is defined as:

$$\mathbf{A}^I \equiv (\mathbf{A}^T\mathbf{A})^{-1}\mathbf{A}^T \tag{1.5.3}$$

The generalized inverse defined in eq.(1.5.3) is referred to as the *Moore-Penrose* generalized inverse, to distinguish it from other inverses. Its existence is guaranteed by the full-rankness assumption of $\mathbf{A}$.

If, in eq.(1.5.1), m is assumed to be smaller than n, but is assumed still to be of full rank, then that equation admits infinitely many solutions. Hence, one particular solution which meets some optimality condition can be chosen. In particular, if a specific $\mathbf{x}_o$ is sought, whose Euclidean norm be a minimum, then $\mathbf{x}_o$ can be expressed as

$$\mathbf{x}_o = \mathbf{A}^\dagger\mathbf{b} \tag{1.5.4}$$

In eq.(1.5.4), $\mathbf{A}^\dagger$ represents another generalized inverse of $\mathbf{A}$, which is defined as:

$$\mathbf{A}^\dagger \equiv \mathbf{A}^T(\mathbf{A}\mathbf{A}^T)^{-1} \tag{1.5.5}$$

Clearly, if $m = n$, then the two foregoing generalized inverses reduce to the inverse of $\mathbf{A}$, the full-rankness assumption reducing to the nonsingularity assumption on $\mathbf{A}$.

An interesting result in connection with $\mathbf{x}_o$ as given by eq.(1.5.2) is that $\mathbf{A}\mathbf{x}_o$ is perpendicular to $\mathbf{e}_o$, the least-square error, which is defined as $\mathbf{b} - \mathbf{A}\mathbf{x}_o$. This is called the *Projection Theorem*. As to the value of $\mathbf{x}_o$ given by eq.(1.5.4), it is pointed out that it lies in the range of $\mathbf{A}^T$, which is natural, because any vector added to it, lying in the nullspace of $\mathbf{A}$, will not modify eq.(1.5.1), and hence the minimum-norm solution to that equation, for $m < n$, is bound to lie in the range of $\mathbf{A}^T$. Hence one can define the following *projector*:

$$\mathbf{B} = \mathbf{1}_n - \mathbf{A}^\dagger\mathbf{A}$$

where $\mathbf{1}_n$ denotes the $n \times n$ identity tensor.

The $n \times n$ tensor $\mathbf{B}$ given above is a projector for it is idempotent, i.e., $\mathbf{B}^2 = \mathbf{B}$. It projects any n-dimensional vector into the nullspace of $\mathbf{A}$, and hence, it is an *orthogonal complement of* $\mathbf{A}$, i.e.,

$$\mathbf{AB} = \mathbf{0}$$

Thus, every n-dimensional vector $\mathbf{y} = \mathbf{Bx}$ lies in the nullspace of $\mathbf{A}$.

Chapter 2 DISPLACEMENT OF A RIGID BODY

2.1 Introduction

Results concerning the motion of a rigid body between two distinct configurations are discussed in this chapter. These configurations are assumed to be *finitely separated*, i.e., the displacements undergone by the points of a bounded subset of the body are assumed to be finite. *Infinitesimally separated* configurations of a rigid body are dealt with in Chapters 3 and 4. The main results of this chapter are *Euler's Theorem*, *Chasles' Theorem*, the characterization of a rigid-body motion through its *screw parameters*, and the Aronhold-Kennedy Theorem. The concepts of *screw* and *pose* of a body are introduced, and some results concerning the displacement field of a rigid-body motion, as well as the compatibility conditions that this field verifies, are derived. Contrary to the common practice, quaternions are deliberately avoided here, the reason for such avoidance being, as explained in Chapter 1, that these require a very special algebra. One aim of this chapter is to show that rotations can be fully studied with linear algebra. The reader interested in quaternions is referred to the original works of Kelland and Tait (1882) and Hamilton (1899). A comprehensive review of the subject is given in (Spring 1986).

2.2 Rotation of a Rigid Body About a Fixed Point

Although rigid-body motions, in general, are nonlinear mappings, a result that was proved in Chapter 1, a class of motions undergone by rigid bodies are linear. Indeed, if a point of a rigid body remains fixed throughout a motion $\mathcal{R}$, then the transformation describing the motion will be shown to be linear.

Let $\mathbf{p}_1$ and $\mathbf{p}_2$ be the position vectors of two points, $\mathcal{P}_1$ and $\mathcal{P}_2$, of a rigid body under a motion leaving fixed a point $\mathcal{O}$ of the body. Moreover, let $\mathbf{p}_1'$ and $\mathbf{p}_2'$ be the images of these vectors under $\mathcal{R}$. It is first shown that $\mathcal{R}$ is additive, next that it is homogeneous, and hence it is linear.

Let

$$\mathbf{e} \equiv \mathcal{R}(\mathbf{p}_1 + \mathbf{p}_2) - (\mathbf{p}_1' + \mathbf{p}_2') \tag{2.2.1a}$$

Taking the Euclidean norm of both sides of eq.(2.2.1a) produces

$$\|\mathbf{e}\|^2 = \|\mathcal{R}(\mathbf{p}_1 + \mathbf{p}_2)\|^2 - 2[\mathcal{R}(\mathbf{p}_1 + \mathbf{p}_2)] \cdot (\mathbf{p}_1' + \mathbf{p}_2') + \|\mathbf{p}_1' + \mathbf{p}_2'\|^2 \tag{2.2.1b}$$

Since the inner product is distributive, the foregoing equation can be expanded as

$$\|\mathbf{e}\|^2 = \|\mathcal{R}(\mathbf{p}_1+\mathbf{p}_2)\|^2 - 2\mathbf{p}_1' \cdot \mathcal{R}(\mathbf{p}_1+\mathbf{p}_2) - 2\mathbf{p}_2' \cdot \mathcal{R}(\mathbf{p}_1+\mathbf{p}_2) + \|\mathbf{p}_1'+\mathbf{p}_2'\|^2 \tag{2.2.1c}$$

Now, since the motion under study is rigid, the magnitude of vectors $\mathbf{p}_i$ and the angle between vectors $\mathbf{p}_i$ and $\mathbf{p}_1 + \mathbf{p}_2$, for $i = 1, 2$, remain unchanged under $\mathcal{R}$, and hence,

$$\|\mathbf{p}_i'\|^2 = \|\mathbf{p}_i\|^2, \quad i = 1, 2 \tag{2.2.2a}$$

and

$$\mathbf{p}_i' \cdot \mathcal{R}(\mathbf{p}_1 + \mathbf{p}_2) = \mathbf{p}_i \cdot (\mathbf{p}_1 + \mathbf{p}_2), \quad i = 1, 2 \tag{2.2.2b}$$

Upon substitution of eqs.(2.2.2a & b) into eq.(2.2.1c), the following is obtained:

$$\|\mathbf{e}\|^2 = -\|\mathbf{p}_1 + \mathbf{p}_2\|^2 + \|\mathbf{p}_1' + \mathbf{p}_2'\|^2$$

If the rigidity of the motion under study is invoked again, it is noted that the two Euclidean norms appearing in the foregoing equation are identical and hence,

$$\|\mathbf{e}\| = 0 \tag{2.2.3}$$

Given the positive-definiteness of the norm, then eq.(2.2.3) implies $\mathbf{e} = \mathbf{0}$, and hence $\mathcal{R}$ is additive.

Now, to prove homogeneity, assume that

$$\mathbf{p}_2 = \alpha \mathbf{p}_1, \quad \alpha > 0 \tag{2.2.4}$$

Then

$$\mathbf{p}_2' = \mathcal{R}(\alpha \mathbf{p}_1) \tag{2.2.5a}$$

From the rigidity assumption,

$$\mathbf{p}_2' \cdot \mathbf{p}_1' = \mathbf{p}_2 \cdot \mathbf{p}_1 \tag{2.2.5b}$$

Substitution of eqs.(2.2.4) and (2.2.5a) into eq.(2.2.5b) yields

$$\mathbf{p}_1' \cdot \mathcal{R}(\alpha \mathbf{p}_1) = \alpha \mathbf{p}_1' \cdot \mathbf{p}_1'$$

or

$$\mathbf{p}_1' \cdot [\mathcal{R}(\alpha \mathbf{p}_1) - \alpha \mathbf{p}_1'] = \mathbf{0} \tag{2.2.5c}$$

Since point P_1 has been assumed arbitrary, eq.(2.2.5c) implies

$$\mathcal{R}(\alpha \mathbf{p}_1) = \alpha \mathcal{R}(\mathbf{p}_1) \tag{2.2.6}$$

thereby proving the linear homogeneity of the operator $\mathcal{R}$ describing the motion under study. Since the operator is also additive, it is linear, *q.e.d.*

From the foregoing discussion, $\mathcal{R}$ can be represented by a *proper orthogonal* dyadic $\mathbf{R}$. Indeed, the rigidity condition implies that the image of $\mathbf{p}$ under $\mathcal{R}$ leaving fixed a point O, denoted by $\mathbf{p}'$, has the same Euclidean norm as $\mathbf{p}$, i.e.,

$$\|\mathbf{p}'\| = \|\mathbf{p}\| \tag{2.2.7}$$

or

$$\mathbf{p}'^T \mathbf{p}' = \mathbf{p}^T \mathbf{p}$$

Since

$$\mathbf{p}' = \mathbf{R}\mathbf{p} \tag{2.2.8}$$

then, eq.(2.2.7) leads to

$$\mathbf{p}^T (\mathbf{R}^T \mathbf{R} - \mathbf{1}) \mathbf{p} = 0 \tag{2.2.9}$$

Clearly, eq.(2.2.9) holds for arbitrary $\mathbf{p}$ if, and only if,

$$\mathbf{R}^T \mathbf{R} = \mathbf{1} \tag{2.2.10}$$

i.e., if $\mathbf{R}$ is orthogonal.

Now, let $\mathbf{p}_1$, $\mathbf{p}_2$, and $\mathbf{p}_3$ denote three unit vectors associated with corresponding points of the rigid body under study. Furthermore, let $\{\mathbf{p}_i\}_1^3$ be an orthonormal set, i.e.,

$$\mathbf{p}_i \cdot \mathbf{p}_j = \delta_{ij} \tag{2.2.11}$$

where δ_{ij} is the *Kronecker delta*. Moreover, $\mathbf{p}_3$ will be assumed to be given by

$$\mathbf{p}_3 = \mathbf{p}_1 \times \mathbf{p}_2 \tag{2.2.12}$$

If $\mathbf{p}_i'$ denotes the image of $\mathbf{p}_i$ under $\mathbf{R}$, then

$$\mathbf{R}\mathbf{p}_3 = \mathbf{R}\mathbf{p}_1 \times \mathbf{R}\mathbf{p}_2 \tag{2.2.13}$$

Moreover,

$$\mathbf{p}_1 \times \mathbf{p}_2 \cdot \mathbf{p}_3 = 1 \tag{2.2.14}$$

and

$$\mathbf{R}\mathbf{p}_1 \times \mathbf{R}\mathbf{p}_2 \cdot \mathbf{R}\mathbf{p}_3 = 1 \tag{2.2.15}$$

Then 3×3 matrices $\mathbf{A}$ and $\mathbf{B}$ can be defined as:

$$\mathbf{A} = [\mathbf{p}_1, \mathbf{p}_2, \mathbf{p}_3], \quad \mathbf{B} = [\mathbf{R}\mathbf{p}_1, \mathbf{R}\mathbf{p}_2, \mathbf{R}\mathbf{p}_3] \tag{2.2.16}$$

Clearly,

$$\mathbf{B} = \mathbf{RA} \tag{2.2.17}$$

From eq.(2.2.14),

$$\det(\mathbf{A}) = 1 \tag{2.2.18}$$

From eqs.(2.2.15 & 2.2.17),

$$\det(\mathbf{R})\det(\mathbf{A}) = 1 \tag{2.2.19}$$

and hence

$$\det(\mathbf{R}) = 1$$

thereby showing that $\mathbf{R}$ is a proper orthogonal dyadic.

2.3 The Theorem of Euler

As a consequence of $\mathbf{R}$ being orthogonal, its proper values are located on the unit circle centered at the origin of the complex plane. Moreover, the fact that $\mathbf{R}$ is proper orthogonal leads to the conclusion that at least one of its proper values is $+1$. Now, let $\mathbf{e}$ be the unit vector associated with this proper value, i.e.,

$$\mathbf{Re} = \mathbf{e} \tag{2.3.1}$$

Moreover, any vector $\mathbf{p} = \alpha\mathbf{e}$, where α is a real number, is left unchanged under $\mathbf{R}$. Hence,

$$\mathbf{Rp} = \mathbf{p} \tag{2.3.2}$$

which states the following result:

Theorem 2.3.1 (Euler 1776): *If a rigid body undergoes a motion leaving fixed one of its points, $\mathcal{O}$, then a set of points of the body, lying on a line that passes through $\mathcal{O}$, remains fixed as well.*

Thus, Euler's Theorem states the existence of an axis of rotation, passing through the fixed point $\mathcal{O}$, in a direction given by $\mathbf{e}$, the proper vector of $\mathbf{R}$ associated with its real proper value $+1$. Such a motion is called a *pure rotation* and is fully defined by *the axis of rotation*, $\mathcal{L}$, and the *angle of rotation*, θ, which is defined next. Let $\mathbf{p}$ be the position vector of an arbitrary point of a rigid body undergoing a pure rotation about $\mathcal{O}$. Moreover, let $\mathbf{e}$ be the unit vector defining the direction of the axis of rotation, the image of $\mathbf{p}$ under $\mathbf{R}$ being denoted by $\mathbf{p}'$. Now vector $\mathbf{p}$ is decomposed into a component along $\mathbf{e}$, $\mathbf{e}(\mathbf{e}\cdot\mathbf{p})$, and a component normal to $\mathbf{e}$, denoted by $\mathbf{r}$. The latter is given as

$$\mathbf{r} \equiv (\mathbf{1} - \mathbf{e}\otimes\mathbf{e})\mathbf{p} \tag{2.3.3a}$$

Hence, **p** can be written as

$$\mathbf{p} \equiv (\mathbf{e} \cdot \mathbf{p})\mathbf{e} + \mathbf{r} \tag{2.3.3b}$$

Thus, $\mathbf{p}'$ can be regarded as the sum of two components, namely, the image of each of the two orthogonal components of **p** appearing in eq.(2.3.3b), under **R**. Moreover, the image of the first component under **R** is that component itself, for it is parallel to the axis of rotation, the image of the second component being denoted by $\mathbf{r}'$. One then has

$$\mathbf{p}' = (\mathbf{e} \cdot \mathbf{p})\mathbf{e} + \mathbf{r}' \tag{2.3.4}$$

Now, the angle of rotation θ is defined in conjunction with the direction of **e**. That is, **e** is not fully defined from the axis of rotation, for a change of sign would leave this vector still parallel to the said axis. In order to destroy this ambiguity, **e** will be defined so that angle θ will lie between 0 and π, when measured along the axis of rotation in the positive direction indicated by **e**. Therefore, the relationship among **r**, $\mathbf{r}'$, **e**, and θ is the following:

$$\mathbf{r} \times \mathbf{r}' = r^2 \sin\theta \mathbf{e} \tag{2.3.5}$$

where r denotes the Euclidean norm of **r**. Moreover, since θ lies between 0 and π, $\sin\theta$ is positive.

From the foregoing discussion, then, given an arbitrary rotation about point $\mathcal{O}$, this is uniquely defined by parameters **e** and θ. Clearly, these are invariant and will be henceforth termed the *natural invariants* of **R**, in order to distinguish them from other invariants, yet to be defined. It will prove useful to group **e** and θ into the 4-dimensional vector ν, defined as

$$\nu = \begin{bmatrix} \mathbf{e} \\ \theta \end{bmatrix} \tag{2.3.6}$$

Vector ν is, thus, an element of $\mathcal{E}^4$, the 4-dimensional Euclidean vector space. Since **e** was defined as a unit vector, ν represents the position vector of a point of the cylinder $\mathbf{e} \cdot \mathbf{e} = 1$, whose axis is the θ axis.

2.4 The Rotation Tensor

In this section an invariant representation of the *rotation tensor* **R** will be derived. This will show **R** as an explicit function of its natural invariants **e** and θ. To this end, $\mathbf{p}'$, as given by eq.(2.3.4), is expressed as a linear transformation of **p**. This is readily accomplished if $\mathbf{r}'$, as appearing in eq.(2.3.4), is written as a linear transformation of **p**. Thus, an expression of $\mathbf{r}'$ in terms of **p** or, correspondingly, of **r**, **e**, and θ is required. This can be derived from eq.(2.3.5), which is now rewritten as follows:

$$\mathbf{1} \times \mathbf{r} \cdot \mathbf{r}' = r^2 \sin\theta \mathbf{e} \tag{2.4.1}$$

However, eq.(2.4.1) cannot be solved for $\mathbf{r}'$ because the coefficient of this is singular. Nevertheless, one can use the following relation, arising from the definition of θ:

$$\mathbf{r}\cdot\mathbf{r}' = r^2\cos\theta \tag{2.4.2}$$

Next, the four scalar equations contained in eqs.(2.4.1) and (2.4.2) can be grouped into one single system of four equations determining $\mathbf{r}'$, namely,

$$\mathbf{A}\mathbf{r}' = \mathbf{b} \tag{2.4.3a}$$

where

$$\mathbf{A} \equiv \begin{bmatrix} \mathbf{1}\times\mathbf{r} \\ \mathbf{r}^T \end{bmatrix}, \quad \mathbf{b} \equiv \begin{bmatrix} r^2\sin\theta\mathbf{e} \\ r^2\cos\theta \end{bmatrix} \tag{2.4.3b}$$

which is a system of the form of that appearing in eq.(1.5.1), with $m = 4$ and $n = 3$. Although such a system is, in general, overdetermined, and admits no solution properly speaking, the system at hand is solvable, for its right-hand side lies in the range of $\mathbf{A}$. The corresponding solution can be readily found resorting to the Moore-Penrose generalized inverse defined in Section 1.5, which yields

$$\mathbf{r}' = \mathbf{A}^I\mathbf{b} \tag{2.4.3c}$$

Moreover, $\mathbf{A}^I$ is readily determined as

$$\mathbf{A}^I = \frac{1}{r^2}\begin{bmatrix} -\mathbf{1}\times\mathbf{r} & \mathbf{r} \end{bmatrix} \tag{2.4.3d}$$

and hence,

$$\mathbf{r}' = \left[\cos\theta(\mathbf{1}-\mathbf{e}\otimes\mathbf{e}) + \sin\theta\mathbf{1}\times\mathbf{e}\right]\mathbf{p} \tag{2.4.3e}$$

Upon substitution of eq.(2.4.3e) into eq.(2.3.4), the following expression for $\mathbf{p}'$ is readily derived:

$$\mathbf{p}' = \left[\mathbf{e}\otimes\mathbf{e} + \cos\theta(\mathbf{1}-\mathbf{e}\otimes\mathbf{e}) + \sin\theta\mathbf{1}\times\mathbf{e}\right]\mathbf{p} \tag{2.4.4a}$$

which is the desired relation, for its shows $\mathbf{p}'$ as a linear transformation of $\mathbf{p}$. The coefficient of $\mathbf{p}$ in eq.(2.4.4a) is defined as the rotation tensor $\mathbf{R}$, i.e.,

$$\mathbf{R} \equiv \mathbf{e}\otimes\mathbf{e} + \cos\theta(\mathbf{1}-\mathbf{e}\otimes\mathbf{e}) + \sin\theta\mathbf{1}\times\mathbf{e} \tag{2.4.4b}$$

Now, the linear invariants of $\mathbf{R}$ are readily obtained as

$$\text{vect}(\mathbf{R}) = \mathbf{e}\sin\theta \tag{2.4.4c}$$

and

$$\text{tr}(\mathbf{R}) = 1 + 2\cos\theta \tag{2.4.4d}$$

Rather than using tr($\mathbf{R}$), it will become apparent that $\cos\theta$ is more convenient. Hence, the following linear invariants of $\mathbf{R}$ are defined:

$$\mathbf{q} = \mathbf{e}\sin\theta \tag{2.4.5a}$$

and

$$q_0 = \cos\theta \tag{2.4.5b}$$

Furthermore, a 4-dimensional vector, λ, will be defined as

$$\lambda = \begin{bmatrix} \mathbf{q} \\ q_o \end{bmatrix} \tag{2.4.6a}$$

Vector λ, thus, defines a rotation uniquely, except for $\theta = \pi$, in which case the information on $\mathbf{e}$ is lost. It is pointed out that the four components of λ are not independent, for they must obey:

$$\lambda \cdot \lambda = 1 \tag{2.4.6b}$$

i.e., λ can be interpreted as the position vector of a point of $\mathcal{E}^4$ lying on the sphere (2.4.6b). It is pointed out, moreover, that, due to the linearity relation between λ and $\mathbf{R}$, and the quadratic constraint upon λ, given in eq.(2.4.6b), it is impossible to obtain a unique triplet of independent invariants defining $\mathbf{R}$. Due to this lack of uniqueness, then, all four scalar invariants contained in vector λ will be carried over, while keeping in mind that these are subject to constraint (2.4.6b). Notice that, when $\theta = \pi$, $\mathbf{R}$ takes on the form:

$$\mathbf{R} = -\mathbf{1} + 2\mathbf{e} \otimes \mathbf{e}, \quad \theta = \pi \tag{2.4.7}$$

Tensor $\mathbf{R}$ in terms of $\mathbf{q}$ and q_0, in general, takes on the form

$$\mathbf{R} = q_0\mathbf{1} + \frac{1 - q_0}{\mathbf{q} \cdot \mathbf{q}}\mathbf{q} \otimes \mathbf{q} + \mathbf{1} \times \mathbf{q}, \quad \mathbf{q} \neq \mathbf{0} \tag{2.4.8a}$$

where the linear dependence of $\mathbf{R}$ upon $\mathbf{q}$ and q_0 is apparent. Indeed, the first term of the foregoing expression is linear in q_0. The second appears quadratic in $\mathbf{q}$ but, since it is divided by the square of the magnitude of $\mathbf{q}$, it can be regarded as linear in q_0. Finally, the third term is linear in $\mathbf{q}$. Thus, the symmetric component of $\mathbf{R}$ can be thought of being linear in q_0, i.e., in $\cos\theta$, which is an even function of θ, whereas the skew-symmetric component of $\mathbf{R}$ is linear in $\mathbf{q}$, i.e., in $\sin\theta$, which is an odd function of θ. An alternative representation of $\mathbf{R}$ in terms of λ can be derived if the identity (2.4.6b) is considered in eq.(2.4.8a). which leads to the following:

$$\mathbf{R} = q_0 + \frac{\mathbf{q} \otimes \mathbf{q}}{1 + q_0} + \mathbf{1} \times \mathbf{q}, \quad q_0 \neq -1 \tag{2.4.8b}$$

Next, the *quadratic invariants* of $\mathbf{R}$, $\mathbf{r}$ and r_0, are defined as:

$$\mathbf{r} = \mathbf{e}\sin(\theta/2), \quad r_0 = \cos(\theta/2) \tag{2.4.9}$$

The foregoing invariants are known as the *Euler parameters* of the ro-tation. They are grouped henceforth in the 4-dimensional vector η, defined as

$$\eta = \begin{bmatrix} \mathbf{r} \\ r_0 \end{bmatrix} \tag{2.4.10a}$$

The four components of η, like those of λ, are not independent, for they obey the constraint

$$\eta \cdot \eta = 1 \tag{2.4.10b}$$

which is, again, a sphere in $\mathcal{E}^4$. The following relationships between the linear and the quadratic invariants are readily derived:

$$\mathbf{q} = 2r_0\mathbf{r}, \quad q_0 = 2r_0^2 - 1 \tag{2.4.10c}$$

$$\mathbf{r} = \frac{\mathbf{q}}{2r_0}, \quad r_0 = \pm\sqrt{\frac{1+q_0}{2}} \tag{2.4.10d}$$

Given the quadratic nature of Euler's parameters, the singularity appearing in connection with λ for $\theta = \pi$ now disappears. In terms of Euler's parameters, the rotation tensor takes on the simple form

$$\mathbf{R} = (2r_0^2 - 1)\mathbf{1} + 2\mathbf{r} \otimes \mathbf{r} + 2r_0\mathbf{1} \times \mathbf{r} \tag{2.4.11}$$

from which the quadratic dependence of $\mathbf{R}$ upon η is apparent.

An additional vector invariant of $\mathbf{R}$, $\mathbf{r}'$, is defined in terms of Euler's parameters as follows:

$$\mathbf{r}' = \mathbf{r}/r_0 \tag{2.4.12}$$

The components of $\mathbf{r}'$ are known as *Rodrigues' parameters.* These present a singularity at $\theta = \pi$, such as the linear invariants, and hence, destroy the advantages of using Euler's parameters.

Now one more invariant representation of $\mathbf{R}$ is derived. To this end, $\mathbf{p}'$, as given by eq.(2.4.3a), is regarded as a function solely of θ, which is true if $\mathbf{e}$ is kept fixed. Next $\mathbf{p}'$ is differentiated with respect to θ and the arising expression is written as a linear transformation of $\mathbf{p}'$, rather than one of $\mathbf{p}$, namely,

$$\frac{d\mathbf{p}'}{d\theta} = \frac{d\mathbf{R}}{d\theta}\mathbf{R}^T\mathbf{p}' \tag{2.4.13}$$

From eq.(2.4.3c),

$$\frac{d\mathbf{R}}{d\theta}\mathbf{R}^T = \mathbf{1} \times \mathbf{e} \tag{2.4.15}$$

and hence,

$$\frac{d\mathbf{p}'}{d\theta} = \mathbf{1} \times \mathbf{e} \cdot \mathbf{p}' \tag{2.4.16}$$

which is a linear stationary differential system, in the sense that the coefficient of $\mathbf{p}'$ is independent of θ, the integration variable. The integral of this system is readily obtained as (Coddington and Levinson 1955):

$$\begin{aligned}\mathbf{p}'(\theta) &= \exp(\mathbf{1} \times \mathbf{e}\theta)\mathbf{p}'(0) \\ &= \exp(\mathbf{1} \times \mathbf{e}\theta)\mathbf{p}\end{aligned} \tag{2.4.17}$$

where, clearly, $\mathbf{p}'(0)$ stands for the value of $\mathbf{p}'$ when $\theta = 0$, which is identical to $\mathbf{p}$, as indicated by eqs.(2.4.3*a* & *b*). Hence, the exponential tensor of eq.(2.4.17) is readily identified with $\mathbf{R}$, i.e.,

$$\mathbf{R} = \exp(\mathbf{1} \times \mathbf{e}\theta) \tag{2.4.18}$$

which is, possibly, the simplest representation of $\mathbf{R}$ in terms of its natural invariants, $\mathbf{e}$ and θ. Application of representation (2.4.18) of the rotation tensor, however, should be done carefully, for it may lead to wrong results. For instance, from the definition of $\mathbf{R}$ in eq.(2.4.3*c*) it follows that rotations do not commute, in general. Thus, if $\mathbf{e}_i$ and θ_i, for $i = 1, \ldots, n$, denote the natural invariants of n independent rotations $\mathbf{R}_1, \ldots, \mathbf{R}_n$, then, in general,

$$\mathbf{R}_i\mathbf{R}_j \neq \mathbf{R}_j\mathbf{R}_i; \quad i, j = 1, \ldots, n; \quad i \neq j \tag{2.4.19}$$

However, from representation (2.4.18), one might be tempted to conclude that

$$\begin{aligned}\mathbf{R}_i\mathbf{R}_j &= e^{\mathbf{1}\times\mathbf{e}_i\theta_i}e^{\mathbf{1}\times\mathbf{e}_j\theta_j} = e^{\mathbf{1}\times(\mathbf{e}_i\theta_i+\mathbf{e}_j\theta_j)}\\ &= e^{\mathbf{1}\times(\mathbf{e}_j\theta_j+\mathbf{e}_i\theta_i)} = \mathbf{R}_j\mathbf{R}_i\end{aligned}$$

which is, clearly, not true, unless $\mathbf{e}_i = \mathbf{e}_j$ or θ_i and θ_j are "small". The first case is obvious; the second is now made apparent. From eq.(2.4.3*c*), for a small number ϵ,

$$\lim_{\theta\to\epsilon} \mathbf{R} = \mathbf{1} + \mathbf{e} \times \mathbf{1}\epsilon \tag{2.4.20}$$

Hence,

$$\begin{aligned}\lim_{\theta_i,\theta_j\to\epsilon} \mathbf{R}_i\mathbf{R}_j &= (\mathbf{1} + \mathbf{e}_i \times \mathbf{1}\epsilon)(\mathbf{1} + \mathbf{e}_j \times \mathbf{1}\epsilon)\\ &= \mathbf{1} + \mathbf{e}_i \times \mathbf{1}\epsilon + \mathbf{e}_j \times \mathbf{1}\epsilon + (\mathbf{e}_i \otimes \mathbf{e}_j - \mathbf{e}_j \otimes \mathbf{e}_i)\epsilon^2\end{aligned}$$

the quadratic term vanishing in the limit, which thus shows that "small" rotations—rotations through "small" angles—do commute.

Next a few additional results concerning the rotation tensor are introduced. As shown in Section 2.2, the rotation tensor is proper orthogonal, one of its proper values thus being $+1$, to which the real proper vector $\mathbf{e}$ is associated. Moreover, the proper values of any $n \times n$ orthogonal tensor $\mathbf{Q}$ lie on the unit circle centered at the origin of the complex plane. Indeed, let μ be a proper value of $\mathbf{Q}$, its associated proper vector being $\mathbf{f}$. Thus,

$$\mathbf{Qf} = \mu\mathbf{f} \tag{2.4.21a}$$

Taking the transpose conjugate of both sides of eq.(2.4.21*a*), one obtains

$$\mathbf{f}^*\mathbf{Q}^* = \overline{\mu}\mathbf{f}^* \tag{2.4.21b}$$

$\overline{\mu}$ standing for the conjugate of μ, whereas the superscript *, introduced in Section 1.2, indicates the transpose conjugate of vectors and matrices. As usual, $\mathbf{f}$ is assumed to be of unit magnitude, i.e.,

$$\mathbf{f}^*\mathbf{f} = 1 \tag{2.4.22}$$

Now multiplying each side of eq.(2.4.21b) times the corresponding side of eq.(2.4.21a) yields

$$\mathbf{f}^*\mathbf{Q}^*\mathbf{Q}\mathbf{f} = \mu\overline{\mu}\mathbf{f}^*\mathbf{f} \tag{2.4.23}$$

If $\mathbf{Q}$ is defined over the real field, of course the following holds:

$$\mathbf{Q}^* = \mathbf{Q}^T \tag{2.4.24}$$

and hence,

$$\mathbf{Q}^*\mathbf{Q} = \mathbf{Q}^T\mathbf{Q} = \mathbf{1} \tag{2.4.25}$$

Upon substitution of eqs.(2.4.22) and (2.4.25) into eq.(2.4.23), the following is obtained:

$$\mu\overline{\mu} \equiv |\mu|^2 = 1 \tag{2.4.26}$$

where $|\cdot|$ denotes the *modulus* of the complex number $(\cdot)$, thereby proving the foregoing assertion.

Hence follows that the three proper values of the rotation tensor $\mathbf{R}$ lie on the unit circle centered at the origin of the complex plane. In fact, its real proper value $+1$ lies on the right-hand side intersection of this circle with the real axis. Now the location of the two remaining proper values of $\mathbf{R}$ is determined. Let these be denoted by μ_1 and μ_2. Hence,

$$\text{tr}(\mathbf{R}) = \mu_1 + \mu_2 + 1 \tag{2.4.27a}$$

$$\det(\mathbf{R}) = \mu_1 \cdot \mu_2 \cdot 1 \tag{2.4.27b}$$

On the other hand, from eq.(2.4.4b),

$$\text{tr}(\mathbf{R}) = 1 + 2\cos\theta \tag{2.4.28a}$$

and, of course,

$$\det(\mathbf{R}) = 1 \tag{2.4.28b}$$

From eqs.(2.4.27a & b) and (2.4.28a & b) it readily follows that

$$\mu_1 = e^{i\theta}, \quad \mu_2 = e^{-i\theta} \tag{2.4.29}$$

The proper vectors of $\mathbf{R}$, $\mathbf{f}$, and $\mathbf{g}$, associated with the proper values μ_1 and μ_2, respectively, are now shown to constitute, with $\mathbf{e}$, an orthonormal set. Indeed, one has

$$\mathbf{f}^*\mathbf{e} = \mathbf{f}^*\mathbf{R}^*\mathbf{R}\mathbf{e} \tag{2.4.30a}$$

Moreover, by asumption,

$$\mathbf{Rf} = e^{i\theta}\mathbf{f}, \quad \mathbf{Re} = \mathbf{e}$$

and hence,

$$\mathbf{f}^*\mathbf{R}^*\mathbf{Re} = e^{-i\theta}\mathbf{f}^*\mathbf{e} \tag{2.4.30b}$$

Upon substitution of eq.(2.4.30*b*) into eq.(2.4.30*a*), one obtains

$$\mathbf{f}^*\mathbf{e} = e^{-i\theta}\mathbf{f}^*\mathbf{e} \tag{2.4.31}$$

However,

$$e^{-i\theta} \neq 0$$

and hence, eq.(2.4.31) holds *for every value of* θ if, and only if,

$$\mathbf{f}^*\mathbf{e} = 0 \tag{2.4.32a}$$

Similarly, the following results can be readily derived:

$$\mathbf{g}^*\mathbf{e} = 0, \quad \mathbf{f}^*\mathbf{g} = 0 \tag{2.4.32b}$$

Moreover, **f** and **g** have been assumed to be normalized, and hence,

$$\|\mathbf{f}\| = \|\mathbf{g}\| = \|\mathbf{e}\| = 1 \tag{2.4.32c}$$

thereby showing that the set $\{\mathbf{e}, \mathbf{f}, \mathbf{g}\}$ is, in fact, orthonormal. Furthermore, the foregoing results hold for arbitrary $n \times n$ orthogonal matrices, whether proper or improper, which is summarizad in the following:

Theorem 2.4.1: *The n proper values of an orthogonal $n \times n$ tensor lie on the unit circle centered at the origin of the complex plane. Moreover, its n proper vectors are mutually orthogonal, and hence form a complete set.*

Additional properties of orthogonal tensors, applicable to rotation tensors, are now presented. From the fact that the determinant of a product of tensors equals the product of their determinants, a proper orthogonal tensor **R** can be factored as the product of two improper orthogonal tensors $\mathbf{H}_1$ and $\mathbf{H}_2$, namely,

$$\mathbf{R} = \mathbf{H}_1\mathbf{H}_2 \tag{2.4.33}$$

Improper orthogonal $n \times n$ tensors represent *reflections* about hyperplanes imbedded in R^n. For example, if $\mathbf{H}_1$ is a reflection about a hyperplane perpendicular to the unit vector $\mathbf{f}_1$, it results to be the following:

$$\mathbf{H}_1 = \mathbf{1} - 2\mathbf{f}_1 \otimes \mathbf{f}_1 \tag{2.4.34a}$$

Similarly, a reflection about a hyperplane perpendicular to the unit vector $\mathbf{f}_2$, say $\mathbf{H}_2$, has the form:

$$\mathbf{H}_2 = \mathbf{1} - 2\mathbf{f}_2 \otimes \mathbf{f}_2 \tag{2.4.34b}$$

The two foregoing tensors, $\mathbf{H}_1$ and $\mathbf{H}_2$, are improper orthogonal, which can be readily proved as follows:

$$\mathbf{H}_1\mathbf{H}_1^T = \mathbf{1} - 4\mathbf{f}_1 \otimes \mathbf{f}_1 + 4\mathbf{f}_1 \otimes \mathbf{f}_1 = \mathbf{1}$$

Moreover,

$$\det(\mathbf{H}_1) = 1 - 2\mathbf{f}_1 \cdot \mathbf{f}_1 = -1$$

thereby completing the intended proof.

Since the product $\mathbf{H}_1\mathbf{H}_2$ is a proper orthogonal tensor, it represents a rotation, say $\mathbf{R}$. This is given, of course, as

$$\mathbf{R} = \mathbf{1} - 2\mathbf{f}_1 \otimes \mathbf{f}_1 - 2\mathbf{f}_2 \otimes \mathbf{f}_2 + 4(\mathbf{f}_1 \cdot \mathbf{f}_2)\mathbf{f}_1 \otimes \mathbf{f}_2 \tag{2.4.35}$$

Now, if the foregoing transformations are defined on the three-dimensional space, then the linear invariants of $\mathbf{R}$, $\cos\theta$ and $\mathbf{e}\sin\theta$, can be readily computed from eq.(2.4.35) as

$$\cos\theta = 2(\mathbf{f}_1 \cdot \mathbf{f}_2)^2 - 1, \quad \mathbf{u}\sin\theta = 4(\mathbf{f}_1 \cdot \mathbf{f}_2)^2\mathbf{f}_1 \times \mathbf{f}_2$$

Equation (2.4.33) shows a factoring of a proper orthogonal tensor into the product of two improper orthogonal tensors, each of which is defined by one n-dimensional unit vector. Such a vector can be regarded as being composed of $n-1$ independent components, and hence a proper orthogonal tensor can be thought of as being defined by $2(n-1)$ independent scalar parameters. Particularly, for the case $n = 3$, the foregoing result might lead one to think that a rotation tensor is constituted by four independent parameters, which is not the case, for such a number has been shown to be three. This point is now clarified upon the derivation of *Cayley's formula*. It applies to any proper orthogonal $n \times n$ tensor and is derived in Bottema and Roth (1979). The said derivation is next paralleled, for quick reference.

Let $\mathbf{Q}$ denote a proper orthogonal transformation of $\mathcal{E}^n$ into itself, which is assumed to have a set of proper values $\{\lambda_i\}_1^n$, none of which is -1. Moreover, let $\mathbf{p}$ and $\mathbf{p}'$ denote the position vector of a point $\mathcal{P}$ in $\mathcal{E}^n$, and its image under $\mathbf{Q}$. Thus,

$$\mathbf{p}' = \mathbf{Q}\mathbf{p} \tag{2.4.36}$$

Since $\mathbf{Q}$ is orthogonal, it preserves the Euclidean norm, i.e.,

$$\|\mathbf{p}'\|^2 = \|\mathbf{p}\|^2$$

which leads to

$$-\mathbf{p}^T\mathbf{p} + (\mathbf{p}')^T\mathbf{p}' = 0$$

or, in factored form, to

$$(\mathbf{p} + \mathbf{p}')^T(\mathbf{p} - \mathbf{p}') = 0 \tag{2.4.37}$$

Now, let

$$\mathbf{a} \equiv \mathbf{p} + \mathbf{p}' = (\mathbf{1} + \mathbf{Q})\mathbf{p}, \quad \mathbf{b} \equiv \mathbf{p} - \mathbf{p}' = (\mathbf{1} - \mathbf{Q})\mathbf{p} \tag{2.4.38}$$

from which $\mathbf{b}$ can be written as

$$\mathbf{b} = (\mathbf{1} - \mathbf{Q})(\mathbf{1} + \mathbf{Q})^{-1}\mathbf{a} \tag{2.4.39}$$

in which $\mathbf{1} + \mathbf{Q}$ is invertible because $\mathbf{Q}$ has been assumed to be proper orthogonal and to have no proper value equal to -1.

Upon substitution of eqs.(2.4.38) and (2.4.39) into eq.(2.4.37), one obtains

$$\mathbf{a}^T(\mathbf{1} - \mathbf{Q})(\mathbf{1} + \mathbf{Q})^{-1}\mathbf{a} = 0 \tag{2.4.40}$$

which is a quadratic form that vanishes for every value of $\mathbf{a}$. Hence, its associated matrix is skew symmetric. Let this matrix be denoted by $\mathbf{S}$, i.e.,

$$\mathbf{S} = (\mathbf{1} - \mathbf{Q})(\mathbf{1} + \mathbf{Q})^{-1} \tag{2.4.41}$$

from which one readily obtains

$$\mathbf{Q}(\mathbf{1} + \mathbf{S}) = \mathbf{1} - \mathbf{S} \tag{2.4.42}$$

Now, one can readily show that both $\mathbf{1} + \mathbf{S}$ and $\mathbf{1} - \mathbf{S}$ are invertible. Indeed, since matrix $\mathbf{S}$ is skew symmetric, $\mathbf{S}^2$ is symmetric and negative definite. Hence, $\mathbf{1} - \mathbf{S}^2$ is symmetric and positive definite, and hence invertible. However, $\mathbf{1} - \mathbf{S}^2$ can be factored as

$$\mathbf{1} - \mathbf{S}^2 = (\mathbf{1} + \mathbf{S})(\mathbf{1} - \mathbf{S}) \tag{2.4.43}$$

Since the left-hand side of eq.(2.4.43) is invertible, each of the factors of its right-hand side is invertible as well. Therefore, eq.(2.4.42) can be solved for $\mathbf{Q}$, thus obtaining

$$\mathbf{Q} = (\mathbf{1} - \mathbf{S})(\mathbf{1} + \mathbf{S})^{-1} \tag{2.4.44}$$

The matrix representation of an $n \times n$ skew-symmetric tensor contains $n(n-1)/2$ independent entries, and hence, an orthogonal tensor is fully defined by the same number of independent scalar parameters. For the case $n = 3$, this number is 3, in accordance with a previous result.

Finally, one more representation of the rotation tensor $\mathbf{R}$ in terms of its natural invariants will be derived. To this end, let

$$\mathbf{E} \equiv \mathbf{e} \times \mathbf{1} \equiv \mathbf{1} \times \mathbf{e} \tag{2.4.45a}$$

Now, from expression (2.4.18), $\mathbf{R}$ can be expanded in a Taylor series of powers of θ as

$$\mathbf{R} = \sum_{0}^{\infty} \mathbf{E}^k \theta^k \tag{2.4.45b}$$

By virtue of the Cayley-Hamilton Theorem (Finkbeiner 1966; Halmos 1974), all powers of $\mathbf{E}$ equal to or greater than 3 can be represented as linear combinations of the first three linearly independent powers of $\mathbf{E}$, namely, $\mathbf{E}^0$, $\mathbf{E}$ and $\mathbf{E}^2$. Hence, three coefficients c_0, c_1 and c_2 can be found so that $\mathbf{R}$ takes on the form

$$\mathbf{R} = c_0\mathbf{E}^0 + c_1\mathbf{E}\theta + c_2\mathbf{E}^2\theta^2 \qquad (2.4.46a)$$

the foregoing coefficients being determined from the three following equations:

$$c_0 + c_1\nu_i + c_2\nu_i^2 = e^{\nu_i}, \quad i = 1, 2, 3 \qquad (2.4.46b)$$

where $\{\nu_i\}_1^3$ is the set of proper values of $\mathbf{E}$. Clearly, the relation between this set and the set of proper values of $\mathbf{R}$, $\{\mu_i\}_1^3$, previously determined as $\{1, e^{i\theta}, e^{-i\theta}\}$, is the following:

$$\mu_i = e^{\nu_i} \qquad (2.4.46c)$$

and hence,

$$\nu_1 = 0, \quad \nu_2 = i\theta, \quad \nu_= - i\theta \qquad (2.4.46d)$$

Upon substitution of the three foregoing values into eqs.(2.4.46b), and solving of the arising equations for the coefficients sought, one obtains

$$c_0 = 1, \quad c_1 = \sin\theta/\theta, \quad c_2 = (1 - \cos\theta)/\theta^2 \qquad (2.4.46e)$$

The desired expression for $\mathbf{R}$ is thus the following:

$$\mathbf{R} = \mathbf{1} + \sin\theta\mathbf{E} + (1 - \cos\theta)\mathbf{E}^2 \qquad (2.4.47)$$

The foregoing representation of $\mathbf{R}$ is identical to that of eq.(2.4.3c). This is made apparent by noting that tensor $\mathbf{E}^2$ is identical to $\mathbf{e} \otimes \mathbf{e} - \mathbf{1}$— see eq.(1.2.9). The results that follow can be derived resorting to *quaternion algebra*, for all the 4-dimensional vectors introduced so far are isomorphic to quaternions. This algebra will not be resorted to, however, for the same results can be derived from linear algebra, which is more widely applicable.

2.5 General Motion of a Rigid Body. The Pose and the Screw of a Rigid Body

The constraint assumed on motions studied in the previous section is now dropped, i.e., it is no longer assumed that one point of the body remains fixed. The main results of this section are Chasles' Theorem and its immediate consequence, namely, that an arbitrary motion of a rigid body is fully described by six independent scalar parameters, which are usually referred to as *the screw parameters* of the motion. It is first assumed that a particular point $\mathcal{A}$, of a body is tracked, i.e., its position vector before and after the body motion has taken place, labelled $\mathbf{a}$ and $\mathbf{a}'$, respectively, are known. Furthermore, let $\mathbf{p}$ and $\mathbf{p}'$ denote the position vector of an arbitrary point $\mathcal{P}$, before and after the motion as well. The displacements of points $\mathcal{A}$ and $\mathcal{P}$ are now denoted by $\mathbf{u}_A$ and $\mathbf{u}_P$, i.e.,

$$\mathbf{u}_A = \mathbf{a}' - \mathbf{a}, \quad \mathbf{u}_P = \mathbf{p}' - \mathbf{p} \tag{2.5.1}$$

A new motion is now defined, producing displacement vectors $\mathbf{u}'_A$ and $\mathbf{u}'_P$. These displacements are defined by

$$\mathbf{u}'_A = \mathbf{u}_A - \mathbf{u}_A = \mathbf{0}, \quad \mathbf{u}'_P = \mathbf{u}_P - \mathbf{u}_A \tag{2.5.2}$$

Under the latter, point $\mathcal{A}$ remains fixed, and hence the motion is a pure rotation about $\mathcal{A}$, defined by a tensor $\mathbf{R}$. Next, $\mathbf{u}'_P$ is written as

$$\mathbf{u}'_P = (\mathbf{p}' - \mathbf{a}') - (\mathbf{p} - \mathbf{a}) \tag{2.5.3}$$

i.e., the new motion can be regarded as the original one under a change of origin, namely to moving point $\mathcal{A}$. Since this is a pure rotation, represented by $\mathbf{R}$, one has the following:

$$\mathbf{p}' - \mathbf{a}' = \mathbf{R}(\mathbf{p} - \mathbf{a})$$

and hence,

$$\mathbf{p}' = \mathbf{a}' + \mathbf{R}(\mathbf{p} - \mathbf{a}) \tag{2.5.4}$$

Thus,

$$\mathbf{u}_P = \mathbf{a}' - \mathbf{p} + \mathbf{R}(\mathbf{p} - \mathbf{a}) = \mathbf{a}' - \mathbf{R}\mathbf{a} + (\mathbf{R} - \mathbf{1})\mathbf{p} \tag{2.5.5}$$

i.e., the displacement of point $\mathcal{P}$ is a linear function of $\mathbf{p}$. It is now pointed out that, in general, no point $\mathcal{P}$ of the body that undergoes a vanishing displacement exists. Indeed, by setting $\mathbf{u}_P$ equal to zero in eq.(2.5.5), one obtains

$$(\mathbf{R} - \mathbf{1})\mathbf{p} = \mathbf{R}\mathbf{a} - \mathbf{a}' \tag{2.5.6a}$$

Tensor $\mathbf{R} - \mathbf{1}$ will appear frequently in what follows. For brevity and for reasons that will become apparent in Chapters 2 and 3 upon defining the angular-velocity and angular-acceleration tensors, the foregoing tensor is defined as the *angular-displacement tensor*, and is henceforth denoted by $\mathbf{D}$. Hence, eq.(2.5.6a) can be rewritten as:

$$\mathbf{D}\mathbf{p} = \mathbf{R}\mathbf{a} - \mathbf{a}' \tag{2.5.6b}$$

Thus, one would be tempted to solve for the position vector $\mathbf{p}$, of a point of zero displacement, from eq.(2.5.6b). However, this cannot be done, for $\mathbf{D}$ is singular, of rank 2, its one-dimensional nullspace being spanned by vector

$\mathbf{e}$, the unit vector parallel to the axis of rotation of $\mathbf{R}$. That is, a vector $\mathbf{p}$ exists that verifies eq.(2.5.6b) if, and only if, $\mathbf{a}' - \mathbf{Ra}$ happens to lie in the range of $\mathbf{D}$. Hence, in general, a vector $\mathbf{f} \neq \mathbf{0}$ exists for arbitrary values of $\mathbf{p}$, that is defined as

$$\mathbf{f} \equiv \mathbf{a}' - \mathbf{Ra} + \mathbf{Dp} \tag{2.5.7}$$

which, by virtue of eq.(2.5.5) is identical to $\mathbf{u}_P$. Thus, even though $\mathbf{f}$ cannot be zeroed in general, it is possible to find a vector $\mathbf{p}_0$ which renders the Euclidean norm of $\mathbf{f}$ a minimum, as discussed in Section 1.5. According with the *Projection Theorem*, mentioned in that section, the vector $\mathbf{p}_0$ minimizing $\|\mathbf{f}\|$ produces an error $\mathbf{f}_0$ which is perpendicular to $\mathbf{Dp}_0$, i.e., which lies in the nullspace of $\mathbf{D}$. Thus, $\mathbf{f}_0$ is parallel to $\mathbf{e}$, and hence, its component perpendicular to $\mathbf{e}$ vanishes, i.e.,

$$(\mathbf{1} - \mathbf{e} \otimes \mathbf{e})\mathbf{f}_0 = \mathbf{0} \tag{2.5.8}$$

Upon substitution of eq.(2.5.7), in terms of $\mathbf{p}_0$, into eq.(2.5.8), the following linear equation in $\mathbf{p}_0$ is obtained:

$$\mathbf{Dp}_0 = \mathbf{Ra} - (\mathbf{1} - \mathbf{e} \otimes \mathbf{e})\mathbf{a} - \mathbf{a}^T\mathbf{ee} \tag{2.5.9a}$$

Again, eq.(2.5.9a) cannot be solved for $\mathbf{p}_0$, for the same reason that it was impossible to solve for $\mathbf{p}$ from eq.(2.5.6b). Nevertheless, since the nullspace of $\mathbf{D}$ is spanned by vector $\mathbf{e}$, its range is of dimension 2, and hence eq.(2.5.9a) contains two linearly independent equations, and thus represents a line parallel to $\mathbf{e}$. Therefore, a set of points $\mathcal{P}_0$ exists, of position vector $\mathbf{p}_0$, whose displacements, represented by $\mathbf{f}_0$, have a minimum Euclidean norm. This set of points is the aforementioned line. One particular point $\overline{\mathcal{P}}_0$, of position vector $\bar{\mathbf{p}}_0$, of this line can be determined, however, by defining it as that whose distance to the origin is a minimum. Hence, $\bar{\mathbf{p}}_0$ is perpendicular to $\mathbf{e}$, i.e.,

$$\mathbf{e}^T\bar{\mathbf{p}}_0 = 0 \tag{2.5.9b}$$

If now eq.(2.5.9a) is rewritten with $\bar{\mathbf{p}}_0$, instead of $\mathbf{p}_0$, equations (2.5.9a & b) represent a formally overdetermined system of four equations with three unknowns. The overdeterminacy of the foregoing system is only formal because the first three equations are compatible with the fourth one. That is, the right-hand side of eq.(2.5.9a) lies in the range of $\mathbf{D}$, whereas eq.(2.5.9b) states that $\bar{\mathbf{p}}_0$ does not lie in the nullspace of $\mathbf{D}$. The foregoing system can thus be represented as

$$\mathbf{A}\bar{\mathbf{p}}_0 = \mathbf{b} \tag{2.5.10a}$$

where $\mathbf{A}$ is a 4×3 matrix and $\mathbf{b}$ is a 4-dimensional vector, both being defined as

$$\mathbf{A} = \begin{bmatrix} \mathbf{D} \\ \mathbf{e}^T \end{bmatrix}, \quad \mathbf{b} = \begin{bmatrix} \mathbf{Ra} - (\mathbf{1} - \mathbf{e} \otimes \mathbf{e})\mathbf{a}' - \mathbf{a}^T\mathbf{e} \otimes \mathbf{e} \\ 0 \end{bmatrix} \tag{2.5.10b}$$

Now $\overline{\mathbf{p}}_0$ can be obtained explicitly in terms of the *Moore-Penrose generalized inverse* of $\mathbf{A}$, $\mathbf{A}^I$ (Golub and Van Loan 1983), as

$$\overline{\mathbf{p}}_0 = \mathbf{A}^I \mathbf{b} \tag{2.5.11a}$$

where $\mathbf{A}^I$ is, from eq.(2.5.10b),

$$\mathbf{A}^T\mathbf{A} = (2)\mathbf{1} + \mathbf{e} \otimes \mathbf{e} - (\mathbf{R} + \mathbf{R}^T) \tag{2.5.11b}$$

Upon substitution of $\mathbf{R}$, as given by eq.(2.4.3c), into eq.(2.5.11b), one obtains

$$\mathbf{A}^T\mathbf{A} = 2(1 - \cos\theta)\mathbf{1} - (1 - 2\cos\theta)\mathbf{e} \otimes \mathbf{e} \tag{2.5.11c}$$

which can be readily inverted as:

$$(\mathbf{A}^T\mathbf{A})^{-1} = \frac{1}{2(1-\cos\theta)}\mathbf{1} + \frac{1-2\cos\theta}{2(1-\cos\theta)}\mathbf{e} \otimes \mathbf{e} \tag{2.5.11d}$$

Hence, from definition (1.5.3),

$$\mathbf{A}^I = [\frac{\mathbf{D}^T}{2(1-\cos\theta)}, \mathbf{e}] \tag{2.5.12}$$

Thus, $\mathbf{p}_0$ is given by:

$$\overline{\mathbf{p}}_0 = \frac{\mathbf{D}^T(\mathbf{R}\mathbf{a} - \mathbf{a}')}{2(1-\cos\theta)} \tag{2.5.13}$$

Substitution of eq.(2.5.13) into eq.(2.5.7) produces the following value $\mathbf{f}_0$ of $\mathbf{f}$:

$$\mathbf{f}_0 = \mathbf{e} \otimes \mathbf{e}(\mathbf{a}' - \mathbf{a}) \tag{2.5.14}$$

Thus, the displacement of minimum Euclidean norm is simply the component of the displacement of any point of the rigid body onto the axis of the accompanying rotation.

The foregoing results are now summarized in the following:

Theorem 2.5.1 (Chasles): *Under the most general motion of a rigid body, a set of points of the body, namely a line parallel to the axis of the rotation involved, undergo a displacement of minimum magnitude that is parallel to that axis. Moreover, the axis passes through a point whose position vector is given by eq.(2.5.13).*

What Chasles' Theorem states is the fact that every rigid-body motion is equivalent to the motion undergone by the nut of a screw, i.e., a rotation about and a translation along the axis of the screw. Hence, a rigid-body motion is also termed a *screw motion*, the parameters that define it, namely, $\theta, \mathbf{e}$ and $\mathbf{p}_0$, being termed, in turn, the *screw parameters* of the motion. It is pointed out that the number of scalar parameters defining a rigid-body motion is seven, but only six of these are independent, for $\mathbf{e}$ obeys the following:

$$\mathbf{e}^T\mathbf{e} = 1 \tag{2.5.15}$$

As a consequence of Chasles' Theorem, one has the following:

Corollary 2.5.1: *The projection of the displacement of all the points of a rigid body onto the axis of its screw is a constant.*

Let u be the projection of the displacement field of a rigid body under general motion onto the screw axis. If θ is the angle of rotation of the associated rotation, then the *pitch* p of the screw motion is defined as:

$$p = \frac{2\pi u}{\theta} \tag{2.5.16}$$

Corollary 2.5.2: *If the displacement of any point of a rigid body is perpendicular to the axis of its screw, then the body undergoes a pure rotation about a point.*

Corollary 2.5.3: *The motion resulting upon subtraction of the displacement of one particular point from the displacement of every point of a rigid body is a pure rotation.*

Corollary 2.5.4: *The difference of the displacements of any two points of a rigid body is perpendicular to its screw axis.*

Now the concepts of *pose* and *screw* of a rigid body can be introduced. The pose of a rigid body is the configuration of the body that is defined *uniquely* by the position vector of one of its points and its orientation with respect to a given reference configuration. From the foregoing it is clear that the pose of a rigid body is totally defined by two vectors, namely, the 3-dimensional position vector $\mathbf{p}$ of a point P of the body, and one of the 4-dimensional vectors ν, λ, or η, defined in eqs.(2.3.7), (2.4.6a), and (2.4.10a), respectively. Thus, the screw of the rigid body under study is defined now as any of the three following forms of the 7-dimensional vector $\mathbf{s}$ given next:

$$\mathbf{s} = \begin{bmatrix} \nu \\ \mathbf{p} \end{bmatrix}, \quad \mathbf{s} = \begin{bmatrix} \lambda \\ \mathbf{p} \end{bmatrix}, \quad \mathbf{s} = \begin{bmatrix} \eta \\ \mathbf{p} \end{bmatrix} \tag{2.5.17}$$

depending on whether the rotation is represented in terms of its natural, its linear, or its quadratic invariants. Of course, in order to avoid ambiguities, the definition being used should be clearly indicated.

2.6 Theorems About the General Motion of a Rigid Body

The motion undergone by a rigid body producing identical displacements of all the points of the body is termed a *pure translation*. Clearly, a motion is a pure translation if, and only if, the arising rotation tensor is the identity. Now, one has the following:

Theorem 2.6.1: *A rigid-body motion is a pure translation if, and only if, the displacements of three noncollinear points of the body are identical.*

Proof: Let P_1, P_2, and P_3 be three noncollinear points of the body, their corresponding position vectors being $\mathbf{p}_1, \mathbf{p}_2$, and $\mathbf{p}_3$. Moreover, let their displacements be $\mathbf{u}_1, \mathbf{u}_2$, and $\mathbf{u}_3$. From eq.(2.5.5), these are:

$$\mathbf{u}_i = \mathbf{a}' - \mathbf{R}\mathbf{a} + \mathbf{D}\mathbf{p}_i, \quad i = 1, 2, 3 \tag{2.6.1}$$

Now, $\mathbf{u}_3$ is subtracted from $\mathbf{u}_1$ and $\mathbf{u}_2$, which produces

$$\mathbf{D}(\mathbf{p}_1 - \mathbf{p}_3) = \mathbf{0}, \quad \mathbf{D}(\mathbf{p}_2 - \mathbf{p}_3) = \mathbf{0} \tag{2.6.2}$$

Next, eqs.(2.6.2) hold if one of the following conditions does: *i*) $\mathbf{p}_1 - \mathbf{p}_3$ and $\mathbf{p}_2 - \mathbf{p}_3$ are parallel to $\mathbf{e}$; *ii*) $\mathbf{D} = \mathbf{0}$. However, the first condition cannot hold, for the three points have been assumed to be noncollinear. Hence, the only possibility is $\mathbf{D} = \mathbf{0}$, i.e., $\mathbf{R} = \mathbf{1}$, which implies that the motion is a pure translation, thereby completing the proof.

Additional theorems regarding the displacement field of a rigid body under general motion follow.

Theorem 2.6.2: *The non-identical displacements of three points of a rigid body are coplanar if, and only if, one of the following three conditions is met:*

i) the motion is a pure rotation;
ii) the motion is general and the three points are collinear; and
iii) the motion is general and the three points are noncollinear, but lie in a plane parallel to the screw axis.

Proof: First *sufficiency* is proved. One considers each of the foregoing possibilities at a time, namely,

i) If the motion is a pure rotation and the origin of coordinates is located on the axis of rotation, the displacement $\mathbf{u}$ of any point of position vector $\mathbf{r}$ is given by $\mathbf{u} = \mathbf{D}\mathbf{r}$, where $\mathbf{D}$ is the associated angular-displacement tensor, and $\mathbf{u}$ lies in the range of $\mathbf{D}$. Since the nullspace of $\mathbf{D}$ is of dimension 1—it is spanned by the unit vector $\mathbf{e}$ defining the direction of the axis of rotation—, then the range of $\mathbf{D}$ is of dimension 2, namely, a plane passing through the origin and perpendicular to the axis of rotation. Thus, all displacements of the body are coplanar.

ii) Let $\mathcal{A}, \mathcal{B}$, and $\mathcal{C}$ be the given three collinear points of the rigid body, **a**, **b**, and **c** being their respective position vectors. Thus, vectors $\mathbf{b}-\mathbf{a}$ and $\mathbf{c}-\mathbf{a}$ are linearly dependent and are related by

$$\mathbf{c}-\mathbf{a}=\alpha(\mathbf{b}-\mathbf{a}) \cdot \tag{2.6.3}$$

where α is a scalar. From eq.(2.5.5), the displacement of $\mathcal{C}$, $\mathbf{u}_C$, can be written as

$$\begin{aligned} \mathbf{u}_C &= \mathbf{a}' + \mathbf{R}(\mathbf{c}-\mathbf{a}) - \mathbf{c} \\ &= \mathbf{a}' - \mathbf{a} + \mathbf{c} + \mathbf{D}(\mathbf{c}-\mathbf{a}) - \mathbf{c} \\ &= \mathbf{u}_A + \alpha\mathbf{D}(\mathbf{b}-\mathbf{a}) \end{aligned} \tag{2.6.4}$$

But, also from eq.(2.5.5),

$$\mathbf{D}(\mathbf{b}-\mathbf{a}) = \mathbf{u}_B - \mathbf{u}_A \tag{2.6.5}$$

Hence, eq.(2.6.4) can be written as

$$\mathbf{u}_C = (1-\alpha)\mathbf{u}_A + \alpha\mathbf{u}_B$$

and it thus becomes apparent that the three displacements are coplanar.

iii) Let $\mathcal{S}$ be a point of the body lying on the screw axis, and **s** its position vector. The displacements of the three given points can be written as (Angeles 1982):

$$\begin{aligned} \mathbf{u}_A &= \mathbf{u}_S + \mathbf{D}(\mathbf{a}-\mathbf{s}) \\ \mathbf{u}_B &= \mathbf{u}_S + \mathbf{D}(\mathbf{b}-\mathbf{s}) \\ \mathbf{u}_C &= \mathbf{u}_S + \mathbf{D}(\mathbf{c}-\mathbf{s}) \end{aligned}$$

Under the assumption that the three given points lie in a plane parallel to the screw axis, vectors $\mathbf{b}-\mathbf{a}$, $\mathbf{c}-\mathbf{a}$, and $\mathbf{u}_S$ are coplanar, and hence they observe the following relation:

$$\mathbf{c}-\mathbf{a} = \alpha(\mathbf{b}-\mathbf{a}) + \beta\mathbf{u}_S$$

or

$$\mathbf{c} = (1-\alpha)\mathbf{a} + \alpha\mathbf{b} + \beta\mathbf{u}_S$$

Substitution of the latter expression into $\mathbf{u}_C$ as given above, after cancellations and rearrangement of terms, produces

$$\mathbf{u}_C = \mathbf{u}_A - \alpha\mathbf{D}(\mathbf{a}-\mathbf{b})$$

But, from the above expressions for $\mathbf{u}_A$ and $\mathbf{u}_B$,

$$\mathbf{u}_A - \mathbf{u}_B = \mathbf{D}(\mathbf{a}-\mathbf{b})$$

Thus, from the latter expression for $\mathbf{u}_C$,

$$\mathbf{u}_C = (1-\alpha)\mathbf{u}_A + \alpha\mathbf{u}_B$$

thereby showing that the three displacements are coplanar.

Next, *necessity* is proved.

If $\mathbf{u}_A$, $\mathbf{u}_B$, and $\mathbf{u}_C$ are assumed to be coplanar, then

$$\det(\mathbf{u}_A, \mathbf{u}_B, \mathbf{u}_C) = 0$$

From eq.(2.5.5), one has

$$\mathbf{u}_B = \mathbf{u}_A + \mathbf{D}(\mathbf{b}-\mathbf{a})$$
$$\mathbf{u}_C = \mathbf{u}_A + \mathbf{D}(\mathbf{c}-\mathbf{a})$$

Hence, the vanishing of the foregoing determinant can be stated, after a few simplifications, as:

$$\det[\,(\mathbf{u}_A, \mathbf{D}(\mathbf{b}-\mathbf{a}), \mathbf{D}(\mathbf{c}-\mathbf{a})\,] = 0$$

or, in Cartesian-vector notation, as

$$\mathbf{D}(\mathbf{b}-\mathbf{a}) \times \mathbf{D}(\mathbf{c}-\mathbf{a}) \cdot \mathbf{u}_A = 0$$

The cross product of the left-hand side of the foregoing equation being the product of two vectors lying in a plane perpendicular to the axis of rotation, can be written in the form

$$\mathbf{D}(\mathbf{b}-\mathbf{a}) \times \mathbf{D}(\mathbf{c}-\mathbf{a}) = \alpha\mathbf{e}$$

where the scalar α can be readily computed by substitution of eq.(2.4.3c) into the above expression, which yields

$$\alpha = 2(1-\cos\theta)\mathbf{e} \times (\mathbf{b}-\mathbf{a}) \cdot (\mathbf{c}-\mathbf{a})$$

$\mathbf{e}$ and θ being the unit vector parallel to the axis of rotation and the angle of rotation, respectively. The double product of interest can thus vanish under any one of the following conditions:

i) $\mathbf{e} \cdot \mathbf{u}_A = 0$, which, from Corollary 2.5.2, states that the motion is a pure rotation.

ii) $\alpha = 0$, which is verified under any of the following conditions:

ii.i) $1-\cos\theta = 0$, which implies $\theta = 0$, i.e., the motion reduces to a pure translation, a case that has been discarded from the outset.

ii.ii) $\mathbf{e} \times (\mathbf{b}-\mathbf{a}) \cdot (\mathbf{c}-\mathbf{a}) = 0$, which is verified if either $(\mathbf{b}-\mathbf{a}) \times (\mathbf{c}-\mathbf{a})$ vanishes, or if the three vectors $\mathbf{e}$, $\mathbf{b}-\mathbf{a}$, and $\mathbf{c}-\mathbf{a}$ are linearly dependent. The first case means that the three given points are collinear, and hence is discarded; the second case means that the three given points lie in a plane parallel to the axis of the rotation involved, i.e., the three points lie in a plane parallel to the screw axis, thereby completing the proof.

As a consequence of the foregoing, one has:

Corollary 2.6.1: *Let $\mathcal{A}, \mathcal{B}$, and $\mathcal{C}$ be three points of a rigid body undergoing an arbitrary motion, and $\mathbf{u}_A, \mathbf{u}_B, \mathbf{u}_C$ the corresponding displacements. The two difference vectors $\mathbf{u}_A - \mathbf{u}_C$ and $\mathbf{u}_B - \mathbf{u}_C$ are parallel if, and only if, the points lie in a plane parallel to the screw axis.*

Corollary 2.6.2: *The displacements of any two points of a rigid body cannot be parallel and differentt, unless the body undergoes a pure rotation.*

Corollary 2.6.3: *If the displacements of any two points of a rigid body are parallel, then either i) the displacements are identical and belong to points lying on a line parallel to the screw axis, or ii) the displacements are different, in which case the motion is a pure rotation and the said two points lie on a line intersecting the axis of rotation.*

Corollary 2.6.4: *Using the notation of Corollary 2.6.1, if $\mathbf{u}_B = \beta \mathbf{u}_A$ and $\mathbf{u}_C = \mathbf{0}$, where β is a nonzero scalar, then the motion is a pure rotation and its axis is parallel to vector $\mathbf{b} - \mathbf{c} - \beta(\mathbf{a} - \mathbf{c})$.*

2.7 Compatibility Equations

From the foregoing it is clear that the rotation $\mathbf{R}$ carrying a rigid body from configuration $\mathcal{C}_0$ to configuration $\mathcal{C}$ can be uniquely determined if the position vectors of three noncollinear points of the body are known both in $\mathcal{C}_0$ and in $\mathcal{C}$. Let $\mathbf{p}_i$ and $\mathbf{p}'_i$, for $i = 1, 2, 3$, denote the position vectors of three noncollinear points of the body in $\mathcal{C}_0$ and in $\mathcal{C}$, respectively. Moreover, let $\mathbf{c}$ denote the position vector of the centroid of the three foregoing points in $\mathcal{C}_0$, $\mathbf{c}'$ denoting that in $\mathcal{C}$, i.e.,

$$\mathbf{c} = \frac{1}{3}\sum_1^3 \mathbf{p}_i, \quad \mathbf{c}' = \frac{1}{3}\sum_1^3 \mathbf{p}'_i \tag{2.7.1}$$

Furthermore, the triads of vectors $\{\rho_i\}_1^3$ and $\{\rho'_i\}_1^3$ are now defined as follows:

$$\rho_i = \mathbf{p}_i - \mathbf{c}, \quad \rho'_i = \mathbf{p}'_i - \mathbf{c}', \quad i = 1, 2, 3 \tag{2.7.2}$$

Rotation $\mathbf{R}$ can now be computed from the sets of vectors of eq.(2.7.2). Indeed, a set of orthonormal vectors $\{\mathbf{e}_1, \mathbf{e}_2\}$ can be computed from $\{\rho_i\}_1^3$ that spans the plane of this set. The images $\{\mathbf{f}_1, \mathbf{f}_2\}$ of $\{\mathbf{e}_1, \mathbf{e}_2\}$ under $\mathbf{R}$

can now be obtained from the second triad $\{\rho_i'\}_1^3$. Finally, vectors $\mathbf{e}_3$ and $\mathbf{f}_3$ can be obtained as

$$\mathbf{e}_3 = \mathbf{e}_1 \times \mathbf{e}_2, \quad \mathbf{f}_3 = \mathbf{f}_1 \times \mathbf{f}_2 \tag{2.7.3}$$

Now, tensors $\mathbf{E}$ and $\mathbf{F}$ are defined as

$$\mathbf{E} = [\mathbf{e}_1, \mathbf{e}_2, \mathbf{e}_3], \quad \mathbf{F} = [\mathbf{f}_1, \mathbf{f}_2, \mathbf{f}_3]$$

Hence,

$$\mathbf{F} = \mathbf{RE} \tag{2.7.4}$$

and the rotation $\mathbf{R}$ can be derived from eq.(2.7.4) as

$$\mathbf{R} = \mathbf{FE}^T \tag{2.7.5}$$

Computational details involved in the derivation of $\mathbf{R}$, as indicated above, are given in Angeles (1986-1). In the foregoing, clearly, vectors ρ_i and ρ_i' should verify the rigidity conditions of the motion involved, i.e.,

$$(\rho_i')^T \rho_j' = \rho_i{}^T \rho_j; \quad i, j = 1, 2, 3 \tag{2.7.6}$$

The foregoing conditions are most readily expressed if tensors $\mathbf{P}$ and $\mathbf{P}'$ are introduced, as defined next:

$$\mathbf{P} = [\, \rho_1, \rho_2, \rho_3 \,], \quad \mathbf{P}' = [\, \rho_1', \rho_2', \rho_3' \,] \tag{2.7.7}$$

Now, eqs.(2.7.6) are equivalent to the following:

$$(\mathbf{P}')^T \mathbf{P}' = \mathbf{P}^T \mathbf{P} \tag{2.7.8}$$

which are the compatibility equations of interest.

Chapter 3 VELOCITY ANALYSIS OF RIGID-BODY MOTIONS

3.1 Introduction

In this chapter the angular velocity of a rigid-body motion is introduced as a skew-symmetric tensor, its linear vector invariant being defined as the *angular-velocity vector* of the given motion. The linear relations between the angular-velocity vector and the time rates of change of the natural, the linear, and the quadratic invariants of the rotation tensor are derived. The relation between the angular-velocity vector and the time-rate of change of the quadratic invariants—Euler's parameters—of the rotation tensor have been reported previously, e.g., in Wittenburg (1977) and Kane, Likins, and Levinson (1983). A comprehensive study of the relations between the first and second time derivatives of the Euler parameters and the angular-velocity and angular-acceleration vectors was reported by Nikravesh, Wehage, and Kwon (1985). Apart from these, the other relations are derived for the first time in invariant form. A preliminary derivation of the relation between the angular-velocity vector and the time rate of change of the linear invariants was first introduced in Angeles (1985). Spring (1986) includes a table showing some of the results that are derived here. Furthermore, a theorem related to the velocity distribution in a rigid body, paralleling that of Chasles' of Chapter 2, is proven. Next, the Theorem of Aronhold-Kennedy, pertaining to the relative motion of three rigid bodies, is proven. Additional theorems related to the velocity distribution throughout a moving rigid body are presented and proven, and the concept of *twist* of a rigid body is introduced. Finally, the problem of determining the angular velocity of a rigid-body motion from point-velocity data is discussed, and compatibility equations which the given data should verify, are derived.

3.2 Motion of a Rigid Body About a Fixed Point

In this Chapter, the rotation tensor $\mathbf{R}$ is assumed to be a continuous and differentiable function of time t, and hence will be represented as $\mathbf{R}(t)$, as needed. Then, $\mathbf{p}'$, as given by eq.(2.4.3*b*), and denoting the position vector of point $\mathcal{P}$ in the rotated configuration of the rigid body, is a continuous and differentiable function of time as well. Upon differentiation of both sides of that equation with respect to time t, one obtains the following expression for the velocity of $\mathcal{P}$, $\mathbf{v}(t)$:

$$\mathbf{v}(t) = \dot{\mathbf{R}}\mathbf{p} \tag{3.2.1}$$

where $\mathbf{p}$ is a constant, for it is the position vector of $\mathcal{P}$ in its original position. If now eq.(2.4.3*b*) is solved for $\mathbf{p}$, eq.(3.2.1) takes on the form:

$$\mathbf{v}(t) = \dot{\mathbf{R}}\mathbf{R}^T(t)\mathbf{p}'(t) \tag{3.2.2a}$$

which is an expression for the velocity of point $\mathcal{P}$ in terms of its *current* position vector. The product $\dot{\mathbf{R}}\mathbf{R}^T$ is the *angular-velocity tensor* of the body, henceforth denoted by $\boldsymbol{\Omega}$, i.e.,

$$\boldsymbol{\Omega}(t) = \dot{\mathbf{R}}(t)\mathbf{R}^T(t) \tag{3.2.2b}$$

Hence, eq.(3.2.2*b*) can be rewritten as:

$$\mathbf{v} = \boldsymbol{\Omega}\mathbf{p}' \tag{3.2.3}$$

where the argument t has been dropped, for all three variables appearing therein are functions of time.

From the definition of $\boldsymbol{\Omega}$, it is apparent that it is skew symmetric. Indeed, since $\mathbf{R}(t)$ is orthogonal, one has

$$\mathbf{R}(t)\mathbf{R}^T(t) = \mathbf{1} \tag{3.2.4a}$$

Upon differentiation of both sides of eq.(3.2.4*a*) with respect to time, one obtains

$$\dot{\mathbf{R}}\mathbf{R}^T + \mathbf{R}\dot{\mathbf{R}}^T = \mathbf{0} \tag{3.2.4b}$$

where the argument t is obvious, and hence is dropped. From eq.(3.2.4*b*) it is apparent that

$$\boldsymbol{\Omega} = -\boldsymbol{\Omega}^T \tag{3.2.4c}$$

Tensor $\boldsymbol{\Omega}$ being skew symmetric, its trace vanishes identically, and thus yields no information on the nature of the motion under study. However, its vector, henceforth denoted by $\omega(t)$ and termed the *angular-velocity vector*, does not vanish identically. It plays an important role in kinematics and is thus formally introduced as

$$\omega = \operatorname{vect}(\boldsymbol{\Omega}) \tag{3.2.5}$$

Thus, eq.(3.2.3) can be rewritten as:

$$\mathbf{v} = \omega \times \mathbf{p}' \tag{3.2.6}$$

Since $\mathbf{\Omega}$ is skew symmetric and of 3×3, it is singular. This follows from the fact that the determinant of every $n \times n$ skew-symmetric matrix vanishes if n is odd. Hence, $\mathbf{\Omega}$ has at least one vanishing proper value. In fact, it has only one vanishing and two complex conjugate proper values, which is a property of skew-symmetric Cartesian tensors. Indeed, let $\mathbf{A}$ be such a tensor, λ being one of its proper values. Its characteristic equation is

$$\det(\lambda \mathbf{1} - \mathbf{A}) = \mathbf{0} \tag{3.2.7}$$

Since transposing a tensor, or a matrix, does not alter its characteristic equation, one has also,

$$\det(\lambda \mathbf{1} - \mathbf{A}^T) = 0 \tag{3.2.8a}$$

which, from the assumption that $\mathbf{A} = -\mathbf{A}^T$, leads to

$$-\det(-\lambda \mathbf{1} - \mathbf{A}) = 0 \tag{3.2.8b}$$

thereby showing that, if λ is a proper value of $\mathbf{A}$, then its negative, -λ, is also a proper value of $\mathbf{A}$. This can only happen if the characteristic equation of $\mathbf{A}$ has the form

$$\lambda(\lambda + \sigma)(\lambda - \sigma) = 0 \tag{3.2.9}$$

where σ is readily derived from the fact that

$$\mathrm{tr}(\mathbf{A}^2) = \lambda_1^2 + \lambda_2^2 + \lambda_3^2 \tag{3.2.10}$$

In eq.(3.2.10), $\{\lambda_i\}_1^3$ is the set of proper values of $\mathbf{A}$, and hence,

$$\mathrm{tr}(\mathbf{A}^2) = 2\sigma^2$$

Thus,

$$\sigma^2 = \frac{1}{2}\mathrm{tr}(\mathbf{A}^2) \tag{3.2.11}$$

However, since $\mathbf{A}^T\mathbf{A}$ is positive definite, $\mathbf{A}^2$ is negative definite. Hence, from eq.(3.2.11), one has

$$\sigma = \pm j\sqrt{-\frac{1}{2}\mathrm{tr}(\mathbf{A}^2)}, \quad j \equiv \sqrt{-1} \tag{3.2.12}$$

The following has been proved:

Theorem 3.2.1: *The characteristic equation of the angular-velocity tensor is*

$$\lambda^3 - \frac{1}{2}\mathrm{tr}(\mathbf{\Omega}^2)\lambda = 0 \tag{3.2.13}$$

and its proper values are

$$\lambda_1 = 0, \quad \lambda_2 = j\sqrt{-\frac{1}{2}\mathrm{tr}(\mathbf{\Omega}^2)}, \quad \lambda_3 = -j\sqrt{-\frac{1}{2}\mathrm{tr}(\mathbf{\Omega}^2)}$$

Moreover, $\mathbf{\Omega}$ *verifies*

$$\mathbf{\Omega}^{2k} = \frac{\mathrm{tr}(\mathbf{\Omega}^2)}{2^{k-1}}\mathbf{\Omega}^2, \quad \mathbf{\Omega}^{2k-1} = \frac{\mathrm{tr}^k(\mathbf{\Omega}^2)}{2^{k-1}}\mathbf{\Omega}, \quad k \geq 2 \tag{3.2.14}$$

Proof: This follows directly from eqs.(3.2.9, 3.2.11, 3.2.12) and the Cayley-Hamilton Theorem (Finkbeiner 1966; Halmos 1974).

The proper vector of $\mathbf{\Omega}$ associated with 0 is, thus, parallel to ω. This vector spans, thus, a one-dimensional space, i.e., a line passing through the fixed point, whose points all have a zero velocity. This line is henceforth referred to as *the instantaneous axis of rotation* of the motion under study.

The relations between the angular-velocity tensor or vector and the natural invariants, $\mathbf{e}$ and θ, and their time derivatives, are next derived. To this end, a few basic results are first introduced.

Henceforth, $\mathbf{S}$ is a skew-symmetric 3×3 tensor.

Theorem 3.2.2: *If*

$$\mathbf{p} = \mathrm{vect}(\mathbf{P})$$

then,

$$-\mathbf{p} = \mathrm{vect}(\mathbf{P}^T)$$

This result follows directly from the definition of $\mathrm{vect}(\cdot)$, and hence its proof is straightforward. Moreover,

Theorem 3.2.3:

$$\mathrm{vect}(\mathbf{SP}) = \frac{1}{2}[\mathrm{tr}(\mathbf{P})\mathbf{1} - \mathbf{P}]\mathrm{vect}(\mathbf{S})$$

Proof: Let $\mathbf{t}$ denote $\mathrm{vect}(\mathbf{SP})$. One then has:

$$t_i = \frac{1}{2}\epsilon_{ijk}s_{kl}p_{lj} \tag{3.2.15}$$

Since $\mathbf{S}$ is skew symmetric, it can be represented in terms of vect($\mathbf{S}$), henceforth denoted by $\mathbf{s}$, as

$$\mathbf{S} = \mathbf{1} \times \mathbf{s} \tag{3.2.16}$$

or, in index notation, as

$$s_{kl} = \epsilon_{lmn}\delta_{km}s_n = \epsilon_{knl}s_n \tag{3.2.17}$$

where ϵ_{lmn} is the Levi-Civita tensor, introduced in eq.(1.2.1).

Upon substitution of eq.(3.2.17) into eq.(3.2.15), and simplification of the arising expression, one obtains

$$t_i = \frac{1}{2}\epsilon_{kij}\epsilon_{knl}p_{lj}s_n \tag{3.2.18}$$

Next, the product of the two Levi-Civita tensors is expanded:

$$\epsilon_{kij}\epsilon_{knl} = \delta_{in}\delta_{jl} - \delta_{jn}\delta_{il} \tag{3.2.19}$$

Now, substitution of eq.(3.2.19) into eq.(3.2.18), yields

$$t_i = \frac{1}{2}(p_{jj}\delta_{in} - p_{in})s_n \tag{3.2.20}$$

which is readily identified as half the ith component of $[\text{tr}(\mathbf{P})\mathbf{1} - \mathbf{P}]\text{vect}(\mathbf{S})$, thereby completing the intended proof.

Theorem 3.2.4:

$$\text{tr}(\mathbf{SP}) = -2[\text{vect}(\mathbf{S})]^T[\text{vect}(\mathbf{P})] \tag{3.2.21}$$

Proof:

$$\text{tr}(\mathbf{SP}) = s_{ij}p_{ji} \tag{3.2.22}$$

Following eq.(3.2.17), one can write

$$s_{ij} = \epsilon_{ikj}s_k \tag{3.2.23}$$

Upon substitution of eq.(3.2.23) into eq.(3.2.22), one obtains

$$\text{tr}(\mathbf{SP}) = \epsilon_{ikj}s_kp_{ji}$$

However, the product $\epsilon_{ikj}p_{ji}$ can be readily recognized as

$$\epsilon_{ikj}p_{ji} = -\epsilon_{kij}p_{ji}$$

i.e., the foregoing expression yields twice the negative of the kth component of vect($\mathbf{P}$), thereby proving the theorem.

Theorem 3.2.5: *If* **P** *is proper orthogonal and the absolute value of its trace is different from unity, then*

$$[\mathrm{tr}(\mathbf{P})\mathbf{1} - \mathbf{P}]^{-1} = \frac{1}{\mathrm{tr}^2(\mathbf{P}) - 1}[\mathbf{P}\mathrm{tr}(\mathbf{P}) + \mathbf{P}^T] \tag{3.2.24}$$

Proof: The proper values of **P** have been proven to be

$$\lambda_1 = e^{j\theta}, \quad \lambda_2 = e^{-j\theta}, \quad \lambda_3 = 1$$

in which θ is the angle of rotation associated with **P**. Denoting the proper values of $\mathbf{T} \equiv \mathrm{tr}(\mathbf{P})\mathbf{1} - \mathbf{P}$ by μ_1, μ_2, and μ_3, one has

$$\mu_1 = 1 + e^{j\theta}, \quad \mu_2 = 1 + e^{-j\theta}, \quad \mu_3 = 2\cos\theta$$

Since the characteristic equation of **T** can be written as

$$(\mu - \mu_1)(\mu - \mu_2)(\mu - \mu_3) = 0 \tag{3.2.25}$$

one readily obtains

$$\mu^3 - 2(1 + 2\cos\theta)\mu^2 + 2(1 + 2\cos\theta)(1 + \cos\theta)\mu - 4\cos\theta(1 + \cos\theta) = 0 \tag{3.2.26a}$$

From eq.(3.2.20), the determinant of **T** can be readily recognized as

$$\det(\mathbf{T}) \equiv \det[\mathrm{tr}(\mathbf{P})\mathbf{1} - \mathbf{P}] = 4\cos\theta(1 + \cos\theta) = \mathrm{tr}^2(\mathbf{P}) - 1 \tag{3.2.26b}$$

from which it is apparent that **T** has an inverse as long as $\mathrm{tr}^2(\mathbf{P})$ is distinct from 1. Now, application of the Cayley-Hamilton Theorem (Finkbeiner 1966; Halmos 1974) to **T** yields

$$\mathbf{T}^3 - 2\mathrm{tr}(\mathbf{P})\mathbf{T}^2 + 2\mathrm{tr}(\mathbf{P})(1 + \cos\theta)\mathbf{T} - [\mathrm{tr}^2(\mathbf{P}) - 1]\mathbf{1} = \mathbf{0} \tag{3.2.27}$$

Next, *if* **T** *is invertible*, one can multiply both sides of eq.(3.2.27) times $\mathbf{T}^{-1}$, thus obtaining from the resulting equation,

$$\mathbf{T}^{-1} = \frac{1}{\mathrm{tr}^2(\mathbf{P}) - 1}[\mathbf{T}^2 - 2\mathrm{tr}(\mathbf{P})\mathbf{T} + 2\mathrm{tr}(\mathbf{P})(1 + \cos\theta)\mathbf{1}] \tag{3.2.28}$$

Substitution of **T**, as defined in terms of **P**, into eq.(3.2.28), yields the following:

$$\mathbf{T}^{-1} = \frac{1}{\mathrm{tr}^2(\mathbf{P}) - 1}\{\mathrm{tr}(\mathbf{P})[2(1 + \cos\theta) - \mathrm{tr}(\mathbf{P})]\mathbf{1} + \mathbf{P}^2\} \tag{3.2.29}$$

Furthermore, the characteristic polynomial of $\mathbf{P}$ can be derived from knowledge of the proper values of $\mathbf{P}$, which were found to be 1, $e^{i\theta}$, and $e^{-i\theta}$—see Section 2.4. Hence, the said polynomial is the following:

$$\begin{aligned} P(\lambda) &= (\lambda - 1)(\lambda - e^{i\theta})(\lambda - e^{-i\theta}) \\ &= \lambda^3 - (1 + 2\cos\theta)\lambda^2 + (1 + 2\cos\theta)\lambda - 1 \\ &= \lambda^3 - \mathrm{tr}(\mathbf{P})\lambda^2 + \mathrm{tr}(\mathbf{P})\lambda - 1 \end{aligned} \tag{3.2.30a}$$

Hence, the Cayley-Hamilton Theorem applied to $\mathbf{P}$ yields:

$$\mathbf{P}^3 - \mathrm{tr}(\mathbf{P})\mathbf{P}^2 + \mathrm{tr}(\mathbf{P})\mathbf{P} - \mathbf{1} = \mathbf{0} \tag{3.2.30b}$$

Next, both sides of eq.(3.2.30b) are multiplied by $\mathbf{P}^T$ and the arising equation is solved for $\mathbf{P}^2$, thereby obtaining

$$\mathbf{P}^2 = \mathrm{tr}(\mathbf{P})\mathbf{P} + \mathbf{P}^T - \mathrm{tr}(\mathbf{P})\mathbf{1} \tag{3.2.30c}$$

Finally, Theorem 3.2.5 is proven by substitution of eq.(3.2.30c) into eq.(3.2.29). Moreover, one has

Theorem 3.2.6: *Let* $\mathbf{R}$ *denote the rotation tensor of a rigid-body motion. The associated angular velocity* ω *is related to the time derivative of* $\mathrm{vect}(\mathbf{R})$ *as follows:*

$$\frac{d}{dt}\mathrm{vect}(\mathbf{R}) = \frac{1}{2}[\mathrm{tr}(\mathbf{R})\mathbf{1} - \mathbf{R}]\omega \tag{3.2.31}$$

Proof: From the assumption of the smoothness of $\mathbf{R}$ and the linearity of the $\mathrm{vect}(\cdot)$ and $d(\cdot)/dt$ operators, these commute, and hence, one can write

$$\frac{d}{dt}\mathrm{vect}(\mathbf{R}) = \mathrm{vect}(\dot{\mathbf{R}})$$

which can be rewritten as

$$\frac{d}{dt}\mathrm{vect}(\mathbf{R}) = \mathrm{vect}(\mathbf{\Omega}\mathbf{R})$$

where $\mathbf{\Omega}$ is the angular-velocity tensor associated with $\mathbf{R}$. Since $\mathbf{\Omega}$ is skew symmetric, Theorem 3.2.2 can be applied to the product $\mathbf{\Omega}\mathbf{R}$, thereby obtaining

$$\frac{d}{dt}\mathrm{vect}(\mathbf{R}) = \frac{1}{2}[\mathrm{tr}(\mathbf{R})\mathbf{1} - \mathbf{R}]\mathrm{vect}(\mathbf{\Omega})$$

the desired proof being completed by noting that $\mathrm{vect}(\mathbf{\Omega}) = \omega$. Now, the following is proven:

Theorem 3.2.7: *Let* $\mathbf{e}$ *and* θ *be the natural invariants of a rigid-body rotation represented by* $\mathbf{R}$. *Then,*

$$\dot{\theta} = \mathbf{e}^T\omega \tag{3.2.32}$$

Proof: This follows directly from Theorem 3.2.3. Indeed, upon differentiation of both sides of eq.(2.4.4*b*) with respect to time, one obtains

$$\mathrm{tr}(\dot{\mathbf{R}}) = -2\dot{\theta}\sin\theta \tag{3.2.33}$$

However,

$$\dot{\mathbf{R}} \equiv \dot{\mathbf{R}}\mathbf{R}^T\mathbf{R} = \mathbf{\Omega}\mathbf{R}$$

Hence,

$$\mathrm{tr}(\mathbf{\Omega}\mathbf{R}) = -2\dot{\theta}\sin\theta \tag{3.2.34a}$$

The application of Theorem 3.2.4 allows one to expand the left-hand side of eq.(3.2.34*a*) as

$$\mathrm{tr}(\mathbf{\Omega}\mathbf{R}) = -2\omega^T\mathbf{e}\sin\theta \tag{3.2.34b}$$

the proof of the proposed theorem now following directly from eqs.(3.2.33) and (3.2.34*b*). Furthermore,

Theorem 3.2.8: *Under the assumptions of Theorem 3.2.6,*

$$\omega \times \mathbf{e} = (\mathbf{1} - \mathbf{R})\dot{\mathbf{e}} \tag{3.2.35}$$

Proof: This follows from the fact that **e** is the proper vector of **R** associated with the proper value +1, i.e.,

$$\mathbf{R}\mathbf{e} = \mathbf{e} \tag{3.2.36a}$$

Upon differentiation of both sides of eq.(3.2.36*a*) with respect to time, one obtains

$$\dot{\mathbf{R}}\mathbf{e} + \mathbf{R}\dot{\mathbf{e}} = \dot{\mathbf{e}} \tag{3.2.36b}$$

Now the proof is completed upon transferring the second term of the left-hand side of the foregoing equation to the right-hand side, and rewriting the first term of the aforementioned side as

$$\dot{\mathbf{R}}\mathbf{e} = \dot{\mathbf{R}}\mathbf{R}^T\mathbf{R}\mathbf{e} = \mathbf{\Omega}\mathbf{R}\mathbf{e} = \omega \times \mathbf{e} \tag{3.2.36c}$$

Theorem 3.2.8, as such, is not very useful, for it does not contain an explicit function of $\dot{\mathbf{e}}$ in terms of ω, or viceversa. In fact, eq.(3.2.35) cannot be solved for $\dot{\mathbf{e}}$, because its tensor coefficient is singular. However, an expression of $\dot{\mathbf{e}}$ in terms of ω can be readily derived. Indeed, in light of eq.(3.2.35) and the following fact:

$$\mathbf{e}^T\dot{\mathbf{e}} = 0 \tag{3.2.37}$$

the Moore-Penrose generalized inverse yields the following:

Theorem 3.2.9: *Under the assumptions of Theorem 3.2.6,*

$$\dot{\mathbf{e}} = \frac{\sin\theta}{2(1-\cos\theta)}\mathbf{e}\times(\omega\times\mathbf{e}) + \frac{1}{2}\omega\times\mathbf{e} \tag{3.2.38}$$

Proof: From eq.(3.2.35) and the identity (3.2.37), the following overdetermined linear system of four equations and three unknowns—the three components of vector $\dot{\mathbf{e}}$—is obtained:

$$\mathbf{A}\dot{\mathbf{e}} = \mathbf{b} \tag{3.2.39a}$$

where

$$\mathbf{A} = \begin{bmatrix}\mathbf{D}\\ \mathbf{e}^T\end{bmatrix}, \quad \mathbf{b} = \begin{bmatrix}\mathbf{e}\times\omega\\ 0\end{bmatrix} \tag{3.2.39b}$$

and $\mathbf{D}$ was defined in Section 2.5 as $\mathbf{R}-\mathbf{1}$, the angular-displacement tensor. However, the overdeterminacy of the foregoing system is only apparent, for $\mathbf{b}$ lies, in fact, in the range of $\mathbf{A}$. Indeed, vector $\omega\times\mathbf{e}$ is perpendicular to ω, and so is every vector that lies in the range of $\mathbf{D}$. Additionally, $\mathbf{e}^T\dot{\mathbf{e}}$ vanishes identically, for $\mathbf{e}$ is of constant magnitude. Hence, the least-square approximation of system (3.2.39) is, in fact, its solution. The said approximation is given by the Moore-Penrose generalized inverse of $\mathbf{A}$, $\mathbf{A}^I$, as defined in eq.(1.5.3), and reproduced here for quick reference as:

$$\mathbf{A}^I = (\mathbf{A}^T\mathbf{A})^{-1}\mathbf{A}^T$$

where

$$\mathbf{A}^T\mathbf{A} = 2(1-\cos\theta)\mathbf{1} - (1-2\cos\theta)\mathbf{e}\otimes\mathbf{e}$$

whose inverse exists as long as $\cos\theta$ is not unity, i.e., if $\mathbf{R}$ does not reduce to the identity tensor. If this is not the case, $\mathbf{A}^T\mathbf{A}$ can be readily inverted as

$$(\mathbf{A}^T\mathbf{A})^{-1} = \frac{1}{2(1-\cos\theta)}\mathbf{1} + \frac{1-2\cos\theta}{2(1-\cos\theta)}\mathbf{e}\otimes\mathbf{e}$$

and hence,

$$\mathbf{A}^I = \frac{1}{2(1-\cos\theta)}[\,\mathbf{D}^T,\ -2\cos\theta\mathbf{e}\,] \tag{3.2.40}$$

Substitution of $\mathbf{b}$, as given by eq.(3.2.39b), and $\mathbf{A}^I$, as given by eq.(3.2.40), into eq.(1.5.2) yields the desired result, and the proof is completed.

Equations (3.2.32) and (3.2.38) provide explicit relations for the time rates of change of the natural invariants, $\dot{\theta}$ and $\dot{\mathbf{e}}$, in terms of θ, $\mathbf{e}$, and ω. An inverse relation, i.e., an expression for ω in terms of $\dot{\theta}$, $\dot{\mathbf{e}}$, θ, and $\mathbf{e}$, is next derived. To this end, eqs.(3.2.32) and (3.2.38) are rewritten as:

$$\mathbf{B}\omega = \dot{\nu} \tag{3.2.41a}$$

with $\mathbf{B}$ and $\dot{\nu}$ defined as

$$\mathbf{B} = \begin{bmatrix} \frac{\sin\theta}{2(1-\cos\theta)}(\mathbf{1} - \mathbf{e}\otimes\mathbf{e}) - \frac{1}{2}\mathbf{1}\times\mathbf{e} \\ \mathbf{e}^T \end{bmatrix}, \quad \dot{\nu} = \begin{bmatrix} \dot{\mathbf{e}} \\ \dot{\theta} \end{bmatrix} \tag{3.2.41b}$$

Application of the Moore-Penrose generalized inverse to eq.(3.2.41*a*) leads to the inverse relation of interest, namely,

$$\omega = \mathbf{B}^I \dot{\nu} \tag{3.2.42a}$$

where

$$\mathbf{B}^I = [\sin\theta(\mathbf{1} - \mathbf{e}\otimes\mathbf{e}) + (1-\cos\theta)\mathbf{1}\times\mathbf{e},\ \mathbf{e}] \tag{3.2.42b}$$

Now, by virtue of eqs.(3.2.42*a* & *b*), together with the fact that $\dot{\mathbf{e}}$ is perpendicular to $\mathbf{e}$, eq.93.2.42*a*) reduces to

$$\omega = \mathbf{N}\dot{\nu} \tag{3.2.43a}$$

with $\mathbf{N}$ defined as

$$\mathbf{N} = [\sin\theta\mathbf{1} + (1-\cos\theta)\mathbf{1}\times\mathbf{e},\ \mathbf{e}] \tag{3.2.43b}$$

Hence, one can write ω in expanded form as

$$\omega = \sin\theta\dot{\mathbf{e}} + (1-\cos\theta)\mathbf{e}\times\dot{\mathbf{e}} + \mathbf{e}\dot{\theta} \tag{3.2.43c}$$

In the foregoing discussion, Theorem 3.2.6 provides a relationship between the time derivative of vect($\mathbf{R}$) and ω which, nevertheless, does not possess a suitable inverse relationship. Indeed, the matrix multiplying ω in eq.(3.2.31) becomes singular rather frequently, namely, whenever the angle of rotation, θ, becomes $\pi/2$, π, or $3\pi/2$. A more convenient relationship can be obtained if Theorem 3.2.6 is combined with Theorem 3.2.7, as indicated below.

From eqs.(3.2.31 & 3.2.32), one can write

$$\mathbf{\Lambda}\omega = \dot{\lambda} \tag{3.2.44a}$$

with

$$\mathbf{\Lambda} = \begin{bmatrix} \frac{1}{2}[\mathrm{tr}(\mathbf{R})\mathbf{1} - \mathbf{R}] \\ -\mathbf{q}^T \end{bmatrix}, \quad \dot{\lambda} = \begin{bmatrix} \dot{\mathbf{q}} \\ \dot{q}_0 \end{bmatrix} \tag{3.2.44b}$$

where, from eqs.(2.4.5*a* & *b*),

$$\dot{\mathbf{q}} = \dot{\mathbf{e}}\sin\theta + \mathbf{e}\dot{\theta}\cos\theta \tag{3.2.44c}$$

$$\dot{q}_0 = -\dot{\theta}\sin\theta \tag{3.2.44d}$$

Introduction of the Moore-Penrose generalized inverse in eq.(3.2.44a) yields the following:

$$\omega = \mathbf{\Lambda}^I \dot{\lambda} \equiv \mathbf{L}\dot{\lambda} \qquad (3.2.45a)$$

where $\mathbf{\Lambda}^I$ has been relabelled as $\mathbf{L}$, which is readily found to be

$$\mathbf{L} = [\mathbf{1} + \frac{\mathbf{1} \times \mathbf{q}}{1 + q_0}, -\frac{\mathbf{q}}{1 + q_0}] \qquad (3.2.45b)$$

and hence,

$$\omega = \dot{\mathbf{q}} + \frac{\mathbf{q} \times \dot{\mathbf{q}}}{1 + q_0} - \frac{\dot{q}_0 \mathbf{q}}{1 - q_0} \qquad (3.2.45c)$$

As opposed to eq.(3.2.43c), eq.(3.2.45c) is not applicable over the whole range of values of θ, for the second and the third term of its right-hand side become undefined at $\theta = \pi$. However, ω is defined in this particular case as well, its value being derived by formally taking the limit of the said terms as $\theta \to \pi$. One thus obtains, by application of *L'Hôpital's rule*,

$$\lim_{\theta \to \pi} \frac{\mathbf{q} \times \dot{\mathbf{q}}}{1 + q_0} = \lim_{\theta \to \pi} \frac{\mathbf{q} \times \ddot{\mathbf{q}}}{\dot{q}_0}$$

which becomes equally undefined at $\theta = \pi$. One more application of *L'Hôpital's rule* yields

$$\lim_{\theta \to \pi} \frac{\mathbf{q} \times \dot{\mathbf{q}}}{1 + q_0} = \lim_{\theta \to \pi} \frac{\dot{\mathbf{q}} \times \ddot{\mathbf{q}} + \mathbf{q} \times \mathbf{q}^{(3)}}{\ddot{q}_0} \qquad (3.2.46)$$

where $\mathbf{q}^{(3)}$ stands for the third time derivative of $\mathbf{q}$, and need not be computed, for it appears multiplied by $\mathbf{q}$, which vanishes when $\theta = \pi$. Now, differentiation of both sides of eqs.(3.2.44c & d) with respect to time yields

$$\ddot{\mathbf{q}} = \ddot{\mathbf{e}} \sin\theta + 2\dot{\mathbf{e}}\dot{\theta}\cos\theta - \mathbf{e}\dot{\theta}^2 \sin\theta + \mathbf{e}\ddot{\theta}\cos\theta \qquad (3.2.47a)$$

$$\ddot{q}_0 = -\dot{\theta}^2 \cos\theta - \ddot{\theta}\sin\theta \qquad (3.2.47b)$$

Substitution of eqs.(3.2.47a & b), evaluated at $\theta = \pi$, in eq.(3.2.46), yields

$$\lim_{\theta \to \pi} \frac{\mathbf{q} \times \dot{\mathbf{q}}}{1 + q_0} = 2\mathbf{e} \times \dot{\mathbf{e}} \qquad (3.2.48a)$$

A similar procedure yields

$$\lim_{\theta \to \pi} \frac{\dot{q}_0 \mathbf{q}}{1 + q_0} = -2\mathbf{e}\dot{\theta} \qquad (3.2.48b)$$

Substitution of eqs.(3.2.48a & b) into eq.(3.2.45c) produces

$$\omega = 2\mathbf{e} \times \dot{\mathbf{e}} + \mathbf{e}\dot{\theta}, \quad \text{for} \quad \theta = \pi \qquad (3.2.49)$$

which coincides with the value produced by eq.(3.2.43c). However, the dependence of ω upon the time derivatives of the linear invariants $\mathbf{q}$ and q_0 disappears when $\theta = \pi$. This dependence can be recovered, however, if higher derivatives of the said linear invariants are introduced. Indeed, from eqs.(3.2.44c & d), one has

$$\dot{\mathbf{q}} = -\mathbf{e}\dot{\theta}, \quad \dot{q}_0 = 0, \quad \text{for} \quad \theta = \pi \tag{3.2.50}$$

Similarly, from eqs.(3.2.47a & b),

$$\ddot{\mathbf{q}} = -2\dot{\mathbf{e}}\dot{\theta} - \mathbf{e}\ddot{\theta}, \quad \ddot{q}_0 = \dot{\theta}^2, \quad \text{for} \quad \theta = \pi \tag{3.2.51}$$

Thus, eq.(3.2.49) can be rewritten as

$$\omega = \frac{\dot{\mathbf{q}} \times \ddot{\mathbf{q}}}{\ddot{q}_0} - \dot{\mathbf{q}}, \quad \text{for} \quad \theta = \pi \tag{3.2.52}$$

which is a relationship between ω and the time derivatives of the linear invariants. This relationship, however, contains second time derivatives of the said invariants, but this should be expected, since the relationship between ω and the first derivatives of the linear invariants disappears at $\theta = \pi$.

Next, relationships between ω and the time derivatives of the Euler parameters are derived. From definitions (2.4.6a & b), (2.4.7), (2.4.10a & b), and (2.4.11a), one can solve for $\dot{\eta}$ in terms of $\dot{\lambda}$, thereby obtaining

$$\dot{\eta} = \mathbf{Q}\dot{\lambda} \tag{3.2.53a}$$

with $\mathbf{Q}$ defined as

$$\mathbf{Q} = \frac{1}{4r_0^2}\begin{bmatrix} 2r_0\mathbf{1} & -\mathbf{r} \\ \mathbf{0}^T & r_0 \end{bmatrix} \tag{3.2.53b}$$

Similarly, from the same equations one can solve for $\dot{\lambda}$ in terms of $\dot{\eta}$, which yields

$$\dot{\lambda} = \mathbf{Q}^{-1}\dot{\eta} \tag{3.2.54a}$$

where $\mathbf{Q}^{-1}$ is given by

$$\mathbf{Q}^{-1} = 2\begin{bmatrix} r_0\mathbf{1} & \mathbf{r} \\ \mathbf{0}^T & 2r_0 \end{bmatrix} \tag{3.2.54b}$$

The foregoing expression for $\mathbf{Q}^{-1}$ can be readily verified to produce $\mathbf{1}$ when multiplied by $\mathbf{Q}$, as given by eq.(3.2.53b).

Further, an expression for ω in terms of the time derivatives of the Euler parameters can be readily derived from eqs.(3.2.45a & b) and (3.2.54a & b), i.e.,

$$\omega = \mathbf{L}\mathbf{Q}^{-1}\dot{\eta} \equiv \mathbf{E}\dot{\eta} \tag{3.2.55a}$$

tensor **E** being readily computed as the product indicated in eq.(3.2.55*a*), namely,

$$\mathbf{E} = 2[\, r_0\mathbf{1} + \mathbf{1} \times \mathbf{r},\ -\mathbf{r}\,] \tag{3.2.55b}$$

a result already derived in (Nikravesh, Wehage and Kwon 1985). The inverse relation of eq.(3.2.55*b*) is now derived by solving eq.(3.2.55*a*) for $\dot{\eta}$. Since the said equation is underdetermined in $\dot{\eta}$, it cannot be inverted, properly speaking, for it admits an infinity of solutions. However, its minimum-norm solution can be readily computed by introduction od the generalized inverse of **E**, $\mathbf{E}^{\dagger}$, as defined in eq.(1.5.5). The said generalized inverse can be proven to be given as

$$\mathbf{E}^{\dagger} = \frac{1}{4}\mathbf{E}^T \tag{3.2.55c}$$

and hence,

$$\dot{\eta} = \frac{1}{4}\mathbf{E}^T\omega \tag{3.2.56}$$

Moreover, relationships among $\dot{\nu}$, $\dot{\lambda}$ and $\dot{\eta}$ can be readily derived, as shown below. Indeed, from eqs.(3.2.44*c* & *d*), one has

$$\dot{\nu} = \mathbf{U}\dot{\lambda} \tag{3.2.57a}$$

where **U** is the 4 × 4 tensor defined as follows:

$$\mathbf{U} \equiv -\frac{1}{\|\mathbf{q}\|^3}\begin{bmatrix} \|\mathbf{q}\|^2\mathbf{1} & -q_0\mathbf{q} \\ \mathbf{0}^T & \|\mathbf{q}\|^2 \end{bmatrix} \tag{3.2.57b}$$

its inverse relationship being derived upon differentiation of eqs.(2.4.5*a* & *b*) with respect to time, namely,

$$\dot{\lambda} = \mathbf{U}^{-1}\dot{\nu} \tag{3.2.58a}$$

with $\mathbf{U}^{-1}$ given by

$$\mathbf{U}^{-1} = \begin{bmatrix} -\sin\theta\mathbf{1} & -\cos\theta\mathbf{e} \\ \mathbf{0}^T & -\sin\theta \end{bmatrix} \tag{3.2.58b}$$

Furthermore, the relationship between $\dot{\eta}$ and $\dot{\nu}$ can be derived upon differentiation of eqs.(2.4.9) with respect to time. This yields

$$\dot{\eta} = \mathbf{V}\dot{\nu} \tag{3.2.59a}$$

with **V** defined as

$$\mathbf{V} \equiv \begin{bmatrix} \sin(\theta/2)\mathbf{1} & \frac{1}{2}\cos(\theta/2)\mathbf{e} \\ \mathbf{0}^T & -\frac{1}{2}\sin(\theta/2) \end{bmatrix} \tag{3.2.59b}$$

The inverse relationship of eq.(3.2.59*b*) can be derived likewise, thereby obtaining

$$\dot{\nu} = \mathbf{V}^{-1}\dot{\eta} \tag{3.2.60a}$$

where $\mathbf{V}^{-1}$ is given by

$$\mathbf{V}^{-1} = \frac{1}{\sin^2(\theta/2)}\begin{bmatrix} \sin(\theta/2)\mathbf{1} & \cos(\theta/2)\mathbf{e} \\ \mathbf{0}^T & -2\sin(\theta/2) \end{bmatrix} \tag{3.2.60b}$$

thereby completing the desired relations.

3.3 General Instantaneous Motion of a Rigid Body. The Twist of a Rigid Body

Now, the instantaneous motion of a rigid body in general is studied; that is, contrary to the assumption of the previous Section, in this Section no point of the rigid body is assumed to remain fixed throughout the motion. Given the definition of rigid body, the velocity of any of its points $\mathcal{P}$ is known if the position vector of $\mathcal{P}$, $\mathbf{p}$, is known, and the following are also known: *i*) the position vector $\mathbf{a}$ of a certain point $\mathcal{A}$, different from $\mathcal{P}$; *ii*) the velocity of $\mathcal{A}$, $\dot{\mathbf{a}}$; and *iii*) the angular velocity ω of the body. Indeed, let $\mathcal{A}_0$ and $\mathcal{P}_0$ be these points in a known reference configuration, their position vectors being $\mathbf{a}_0$ and $\mathbf{p}_0$, respectively. Clearly, the differences $\mathbf{p} - \mathbf{a}$ and $\mathbf{p}_0 - \mathbf{a}_0$ are related via the rotation tensor $\mathbf{R}$ in the form

$$\mathbf{p}(t) - \mathbf{a}(t) = \mathbf{R}(t)(\mathbf{p}_0 - \mathbf{a}_0) \tag{3.3.1}$$

where time-varying quantities have been shown explicitly.

Upon differentiation of both sides of eq.(3.3.1) with respect to time, one obtains

$$\dot{\mathbf{p}}(t) - \dot{\mathbf{a}}(t) = \dot{\mathbf{R}}(t)(\mathbf{p}_0 - \mathbf{a}_0) \tag{3.3.2}$$

Solving for $\mathbf{p}_0 - \mathbf{a}_0$ from eq.(3.3.1), and substituting the arising expression into eq.(3.3.2), yields

$$\dot{\mathbf{p}}(t) - \dot{\mathbf{a}}(t) = \dot{\mathbf{R}}(t)\mathbf{R}^T(t)[\mathbf{p}(t) - \mathbf{a}(t)]$$

or, rearranging terms, identifying the product $\dot{\mathbf{R}}\mathbf{R}^T(t)$ as the angular-velocity tensor $\mathbf{\Omega}(t)$, and dropping the argument, for now all quantities are time varying, the latter equation produces

$$\dot{\mathbf{p}} = \dot{\mathbf{a}} + \mathbf{\Omega}(\mathbf{p} - \mathbf{a}) \tag{3.3.3a}$$

which is the relation sought. In that equation, the second term of the right-hand side is referred to as *the relative velocity of $\mathcal{P}$ with respect to $\mathcal{A}$*, its physical interpretation being that, if point $\mathcal{A}$ were fixed, it would yield the velocity of $\mathcal{P}$. Now, one has the following:

Theorem 3.3.1: *The velocity of $\mathcal{P}$ with respect to $\mathcal{A}$ is perpendicular to line $\mathcal{AP}$.*

Proof: This follows immediately from the fact that $\mathbf{\Omega}$ is skew symmetric, which thus produces a vanishing quadratic form $(\mathbf{p} - \mathbf{a})^T\mathbf{\Omega}(\mathbf{p} - \mathbf{a})$.

It is pointed out that, if eq.(3.3.3*a*) is rewritten in terms of the angular-velocity vector ω, namely, as

$$\dot{\mathbf{p}} = \dot{\mathbf{a}} + \omega \times (\mathbf{p} - \mathbf{a}) \tag{3.3.3b}$$

then the foregoing theorem becomes apparent from the vanishing of the product $(\mathbf{p}-\mathbf{a})\cdot\omega\times(\mathbf{p}-\mathbf{a})$.

It is clear, then, that the instantaneous motion of a rigid body is known if the position and the velocity of one of its points, as well as its angular velocity, are known. Thus, one could describe this motion in terms of the foregoing quantities, except that the base-point $\mathcal{A}$ is not unique. One would like to have an invariant description of this motion, instead. This is readily obtained if the motion at hand is described via its *instantaneous-screw parameters*, a subject that is discussed in what follows.

The *instantaneous-screw parameters* consist basically of a line of the body, $\mathcal{L}$, and ω. The line, in turn, will be shown to be the locus of all points of the body moving with one and the same velocity, which is, furthermore, of minimum Euclidean norm. Thus, $\mathcal{L}$ is defined via the minimization of the Euclidean magnitude of $\dot{\mathbf{p}}$, with $\dot{\mathbf{p}}$ given by eq.(3.3.3a). In order to find the set of points of minimum-norm velocity, define a function $z(\mathbf{p})$, that represents one half of the Euclidean norm of that velocity, and that is to be minimized by a proper choice of $\mathbf{p}$, i.,e.,

$$z(\mathbf{p})\equiv\frac{1}{2}\dot{\mathbf{p}}^T\dot{\mathbf{p}}\rightarrow\min_{\mathbf{p}} \tag{3.3.4}$$

Function z is positive definite and quadratic in $\dot{\mathbf{p}}$. Hence, it possesses one unique minimum value, which is attained, as will be shown presently, not at one single value of $\mathbf{p}$, but rather at a set of values of $\mathbf{p}$, defining a line $\mathcal{L}$. Furthermore, the minimum value of z is not zero, in general, for this would require the vanishing of $\dot{\mathbf{p}}$ which, in view of eq.(3.3.3a), would lead to

$$\mathbf{\Omega}\mathbf{p}=\mathbf{\Omega}\mathbf{a}-\dot{\mathbf{a}} \tag{3.3.5}$$

However, eq.(3.3.5) cannot be solved for $\mathbf{p}$, for rank$(\mathbf{\Omega})=2$. This means that, if $\mathbf{\Omega}\mathbf{a}-\dot{\mathbf{a}}$ has a component lying in the nullspace of $\mathbf{\Omega}$, i.e., parallel to ω, then no vector $\mathbf{p}$ exists that verifies eq.(3.3.5). If, on the other hand, such a component of the right-hand side of eq.(3.3.5) is removed from this equation, then not only one vector $\mathbf{p}$, but infinitely many such vectors exist, which verify the resulting equation. This equation takes on the form:

$$\mathbf{\Omega}\mathbf{p}=(\mathbf{1}-\omega\otimes\omega/\|\omega\|^2)(\mathbf{\Omega}\mathbf{a}-\dot{\mathbf{a}}) \tag{3.3.6}$$

Now, eq.(3.3.6) represents a system of three equations—out of which only two are linearly independent—in three unknowns, namely, the components of $\mathbf{p}$. Thus, eq.(3.3.6) has infinitely many solutions, all those differing by a component parallel to ω. If $\mathbf{p}_0$ is a solution of eq.(3.3.6), then the general solution $\mathbf{p}$ can be represented as

$$\mathbf{p}=\mathbf{p}_0+\alpha\omega \tag{3.3.7}$$

where α is any real scalar.

Equation (3.3.7) represents a line parallel to ω passing through a point P_0, of position vector $\mathbf{p}_0$. This line is called *the axis of the instantaneous screw* of the motion. It is fully defined by ω and $\mathbf{p}_0$. Now, point P_0 is uniquely defined if chosen as the point of the axis lying closest to the origin. In this case, $\mathbf{p}_0$ is clearly perpendicular to the axis, and hence, to ω, a condition that can be expressed as

$$\omega^T \mathbf{p}_0 = 0 \tag{3.3.8}$$

Equations (3.3.6) and (3.3.8) are now grouped into one single system of 4 equations and 3 unknowns of the form:

$$\mathbf{A}\mathbf{p}_0 = \mathbf{b} \tag{3.3.9a}$$

where $\mathbf{A}$ is the 4×3 matrix and $\mathbf{b}$ is the 4-dimensional vector that follow:

$$\mathbf{A} = \begin{bmatrix} \mathbf{\Omega} \\ \omega^T \end{bmatrix}, \quad \mathbf{b} = \begin{bmatrix} (\mathbf{1} - \omega \otimes \omega / \|\omega\|^2)(\mathbf{\Omega a} - \dot{\mathbf{a}}) \\ \mathbf{0} \end{bmatrix} \tag{3.3.9b}$$

Thus, eq.(3.3.9a) represents a *formally* overdetermined system of 4 equations in 3 unknowns. In fact, the first three equations of system (3.3.9b) state that $\mathbf{b}$ lies in the range of $\mathbf{\Omega}$, whereas the fourth one states that $\mathbf{p}_0$ is not in the nullspace of $\mathbf{\Omega}$, two facts which are consistent with each other. Furthermore, vector $\mathbf{p}_0$ can be computed as the least-square approximation of eqs.(3.3.9b) in terms of the Moore-Penrose generalized inverse of $\mathbf{A}$, $\mathbf{A}^I$. This can be readily computed by noticing that

$$\mathbf{A}^T \mathbf{A} = \|\omega\|^2 \mathbf{1} \tag{3.3.10a}$$

and hence,

$$\mathbf{A}^I = \frac{1}{\|\omega\|^2}[-\mathbf{\Omega}, \ \omega] \tag{3.3.10b}$$

Thus, $\mathbf{p}_0$ is given explicitly by

$$\mathbf{p}_0 = \frac{1}{\|\omega\|^2}\omega \times (\dot{\mathbf{a}} - \omega \times \mathbf{a}) \tag{3.3.11}$$

One has then proven a result similar to Chasles' Theorem, namely,

Theorem 3.3.2: *The locus of points of a rigid body having a velocity of minimum Euclidean norm is a line parallel to the angular-velocity vector that passes through a point P_0 whose position vector is given by eq.(3.3.11).*

It is pointed out that $\mathbf{p}_0$ and ω uniquely define the Plücker coordinates (Bottema and Roth 1979) of the instantaneous-screw axis (ISA) of the body. Now, as a direct consequence of the foregoing theorem, one has:

Corollary 3.3.1: *The projection of the velocity of any point of a rigid body onto the ISA is a constant, u, given by*

$$u = \dot{\mathbf{a}}^T \omega / \|\omega\| \tag{3.3.12}$$

In eq.(3.3.12), $\dot{\mathbf{a}}$ is the velocity of an arbitrary point $\mathcal{A}$ of the body, which is known, its position vector, $\mathbf{a}$, being also known. Projection u is referred to as the *sliding* of the instantaneous screw, the *pitch* σ of the screw being given by the ratio

$$\sigma = 2\pi u / \|\omega\| \tag{3.3.13}$$

Clearly, if the sliding or the pitch of an instantaneous screw vanishes and $\omega \neq \mathbf{0}$, then the body undergoes a pure rotation.

A result that is closely related to Theorem 3.3.2 is known as the *Aronhold-Kennedy Theorem*, which is next proven. Let $\mathcal{A}$, $\mathcal{B}$, and $\mathcal{C}$ be three rigid bodies undergoing motions with angular-velocity tensors $\mathbf{\Omega}_A$, $\mathbf{\Omega}_B$, and $\mathbf{\Omega}_C$, respectively. Then, three relative instantaneous screws can be defined, namely, those of $\mathcal{B}$ with respect to $\mathcal{A}$, of $\mathcal{C}$ with respect to $\mathcal{A}$, and of $\mathcal{B}$ with respect to $\mathcal{C}$. Moreover, let $\mathbf{v}_{B/A}$ be the velocity of minimum Euclidean norm of the motion of $\mathcal{B}$ with respect to $\mathcal{A}$ which, by virtue of Theorem 3.3.2, is parallel to the ISA of $\mathcal{B}$ with respect to $\mathcal{A}$. Furthermore, let $\mathbf{\Omega}_{B/A}$ denote the angular-velocity tensor of $\mathcal{B}$ with respect to $\mathcal{A}$, the corresponding angular-velocity vector being $\omega_{B/A}$. Velocities $\mathbf{v}_{C/A}$ and $\mathbf{v}_{B/A}$, and tensors $\mathbf{\Omega}_{C/A}$ and $\mathbf{\Omega}_{B/C}$, as well as their associated vectors $\omega_{C/A}$ and $\omega_{B/C}$, are defined correspondingly. Next, let $\mathbf{p}$ denote the vector connecting points $\mathcal{O}$ and $\mathcal{P}$ of the ISA of $\mathcal{B}$ with respect to $\mathcal{A}$ and of $\mathcal{C}$ with respect to $\mathcal{A}$, respectively, directed from the former to the latter, segment $\mathcal{OP}$ belonging to the common perpendicular of these axes. It will be shown that the ISA of $\mathcal{B}$ with respect to $\mathcal{C}$ intersects line $\mathcal{OP}$—not necessarily between points $\mathcal{O}$ and $\mathcal{P}$--and is perpendicular to it, a result that summarizes the Aronhold-Kennedy Theorem.

In order to prove the theorem under study, and without loss of generality, let the origin of $\mathcal{E}^3$ be located at point $\mathcal{O}$, and $\mathbf{r}$ be the position vector of a point $\mathcal{R}$ of the ISA of $\mathcal{B}$ with respect to $\mathcal{C}$. Thus, $\mathcal{R}$ can be regarded as a point of $\mathcal{B}$ or as a point of $\mathcal{C}$. In order to make the distinction apparent, $\mathcal{R}$ will be subscripted as $\mathcal{R}_B$ or $\mathcal{R}_C$, correspondingly. Now, $\mathcal{R}$ or, equivalently, $\mathcal{R}_B$ or $\mathcal{R}_C$, is a point of the ISA of $\mathcal{B}$ with respect to $\mathcal{C}$ if the difference of the velocities of $\mathcal{R}_B$ and $\mathcal{R}_C$ has a minimum Euclidean norm. Let the said velocities be denoted by $\mathbf{v}_{RB}$ and $\mathbf{v}_{RC}$, their difference being labelled $\mathbf{v}$, i.e.,

$$\mathbf{v} = \mathbf{v}_{RB} - \mathbf{v}_{RC} \tag{3.3.14}$$

Now, regarding $\mathcal{O}$ as a point of $\mathcal{B}$, and $\mathcal{P}$ as one of $\mathcal{C}$, $\mathbf{v}_{RB}$ and $\mathbf{v}_{RC}$ can be written as

$$\mathbf{v}_{RB} = \mathbf{v}_O + \mathbf{\Omega}_B \mathbf{r} \tag{3.3.15a}$$

$$\mathbf{v}_{RC} = \mathbf{v}_P + \mathbf{\Omega}_C(\mathbf{r} - \mathbf{p}) \tag{3.3.15b}$$

where, obviously, $\mathbf{v}_O$ and $\mathbf{v}_P$ denote the velocities of $\mathcal{O}$ and $\mathcal{P}$, respectively. Hence,

$$\mathbf{v} = \mathbf{v}_{O/P} + \mathbf{\Omega}_{B/C}\mathbf{r} + \mathbf{\Omega}_C \mathbf{p} \tag{3.3.16a}$$

where

$$\mathbf{v}_{O/P} \equiv \mathbf{v}_O - \mathbf{v}_P \tag{3.3.16b}$$

$$\mathbf{\Omega}_{B/C} \equiv \mathbf{\Omega}_B - \mathbf{\Omega}_C \tag{3.3.16c}$$

In general, no vector $\mathbf{r}$ exists that renders $\mathbf{v}$ zero, for the range of $\mathbf{\Omega}_{B/C}$ is 2, and $\mathbf{v}$ lies in $\mathcal{E}^3$. However, the component of $\mathbf{v}$ lying in the nullspace of $\mathbf{\Omega}_{B/C}$ can be rendered zero by a proper choice of vector $\mathbf{r}$. This is achieved by equating $\mathbf{\Omega}_{B/C}\mathbf{r}$ to the projection of the negative of $\mathbf{v}_{O/P} + \mathbf{\Omega}_C\mathbf{p}$ onto the range of $\mathbf{\Omega}_{B/C}$, i.e., by letting

$$\mathbf{\Omega}_{B/C}\mathbf{r} = -(\mathbf{1} - \omega_{B/C} \otimes \omega_{B/C}/\|\omega_{B/C}\|^2)(\mathbf{v}_{O/P} + \mathbf{\Omega}_C\mathbf{p}) \tag{3.3.17a}$$

However, eq.(3.3.17a) cannot be solved for vector $\mathbf{r}$, for $\mathbf{\Omega}_{B/C}$ is not invertible. In fact, the aforementioned equation defines not one single point, but rather a set of points lying on a line parallel to the vector of $\mathbf{\Omega}_{B/C}$, $\omega_{B/C}$, this line being the ISA of $\mathcal{B}$ with respect to $\mathcal{C}$. Thus, one particular vector, $\mathbf{r}_0$, verifying eq.(3.3.17a), can be defined by specifying that it be of minimum Euclidean norm. Since such a vector is perpendicular to the axis sought, one has

$$\omega_{B/C}^T \mathbf{r}_0 = 0 \tag{3.3.17b}$$

Now, eq.(3.3.17a), written for $\mathbf{r}_0$, and eq.(3.3.17b), define an overdetermined system of four linear equations in three unknowns, of the form

$$\mathbf{A}\mathbf{r}_0 = \mathbf{b} \tag{3.3.18a}$$

where $\mathbf{A}$ and $\mathbf{b}$ are defined as follows:

$$\mathbf{A} = \begin{bmatrix} \mathbf{\Omega}_{B/C} \\ \omega_{B/C}^T \end{bmatrix} \tag{3.18b}$$

$$\mathbf{b} = \begin{bmatrix} -(\mathbf{1} - \omega_{B/C} \otimes \omega_{B/C}/\|\omega_{B/C}\|^2)(\mathbf{v}_{O/P} + \mathbf{\Omega}_C\mathbf{p}) \\ 0 \end{bmatrix} \tag{3.18c}$$

Such as in the case of system (3.3.9*a*), system (3.3.18*a*) is only formally overdetermined, for its right-hand side lies in the range of **A**. Thus, the least-square approximation of that system is, in fact, its solution, which takes on a form similar to that appearing in eq.(3.3.11) for system (3.3.9*a*), namely,

$$\mathbf{r}_0 = \frac{1}{\|\omega_{B/C}\|^2}\omega_{B/C} \times (\mathbf{v}_{O/P} + \omega_C \times \mathbf{p}) \tag{3.3.19a}$$

However,

$$\begin{aligned}\omega_{B/C} \times \mathbf{v}_{O/P} &\equiv (\omega_{B/A} - \omega_{C/A}) \times (\mathbf{v}_O - \mathbf{v}_P) \\ &= \omega_{B/A} \times \mathbf{v}_O - \omega_{B/A} \times \mathbf{v}_P - \omega_{C/A} \times \mathbf{v}_O + \omega_{C/A} \times \mathbf{v}_P\end{aligned}$$

where the first and the fourth terms of the rightmost-hand side vanish because $\mathbf{v}_O$ and $\mathbf{v}_P$ are, by definition, the minimum-norm velocities of $\mathcal{B}$ and $\mathcal{C}$ with respect to $\mathcal{A}$. Thus,

$$\omega_{B/C} \times \mathbf{v}_{O/P} = -(\omega_{B/A} \times \mathbf{v}_P + \omega_{C/A} \times \mathbf{v}_O) \tag{3.3.19b}$$

Moreover,

$$\omega_{B/C} \times (\omega_C \times \mathbf{p}) = (\omega_{B/C} \cdot \mathbf{p})\omega_C - (\omega_{B/C} \cdot \omega_C)\mathbf{p} \tag{3.3.19c}$$

where the first scalar product can be written as

$$\begin{aligned}\omega_{B/C} \cdot \mathbf{p} &= (\omega_{B/A} - \omega_{C/A}) \cdot \mathbf{p} \\ &= \omega_{B/A} \cdot \mathbf{p} - \omega_{C/A} \cdot \mathbf{p} = 0\end{aligned} \tag{3.3.19d}$$

each scalar product vanishing by virtue of the definition of **p**, namely, its perpendicularity to the ISAs of $\mathcal{B}$ with respect to $\mathcal{A}$, and of $\mathcal{C}$ with respect to $\mathcal{A}$. This means that **p** is perpendicular to both $\omega_{B/A}$ and $\omega_{C/A}$.

Upon substitution of eqs.(3.3.19*b*–*d*) into eq.(3.3.19*a*), one obtains

$$\mathbf{r}_0 = -\frac{1}{\|\omega_{B/C}\|^2}[\omega_{B/A} \times \mathbf{v}_P + \omega_{C/A} \times \mathbf{v}_O + (\omega_{B/C} \cdot \omega_C)\mathbf{p}] \tag{3.3.20}$$

The cross products appearing in the right-hand side of eq.(3.3.20) are perpendicular to the ISA of $\mathcal{B}$ with respect to $\mathcal{A}$ and of $\mathcal{C}$ with respect to $\mathcal{A}$, and hence they are parallel to **p**, thereby showing that the ISA of $\mathcal{B}$ with respect to $\mathcal{C}$ intersects the common perpendicular to the other two ISAs. In summary,

Theorem 3.3.3 (Aronhold-Kennedy): *Given three rigid bodies, $\mathcal{A}$, $\mathcal{B}$, and $\mathcal{C}$, undergoing arbitrary motions, the three arising ISAs intersect a common line at right angles.*

Now the concept of *twist* of a rigid body can be introduced. When the pose of a rigid body changes continuously and smoothly with time, its associated screw can be regarded as a continuous and differentiable function of time. The rate at which the screw changes is associated with a quantity known as the twist of the body; which is defined as the following 6-dimensional vector $\mathbf{t}$:

$$\mathbf{t} = \begin{bmatrix} \omega \\ \mathbf{v}_P \end{bmatrix} \tag{3.3.21}$$

where $\mathbf{v}_P$ is the velocity of a point $\mathcal{P}$ of the body, which was used to define its screw, and ω is the angular velocity of the body. From the results of Section 3.2 it is apparent that $\mathbf{t}$ can be represented as a linear transformation of $\dot{\mathbf{s}}$. Indeed, if $\mathbf{s}$ is defined in terms of the natural invariants of the associated rotation, then, one has

$$\mathbf{t} = \begin{bmatrix} \mathbf{N} & \mathbf{0}_{33} \\ \mathbf{0}_{34} & \mathbf{1} \end{bmatrix} \dot{\mathbf{s}} \tag{3.3.22a}$$

where the 3×4 tensor $\mathbf{N}$ was defined in eq.(3.2.43*b*), $\mathbf{0}_{33}$ represents the 3×3 zero tensor, $\mathbf{0}_{34}$ the 3×4 zero tensor and $\mathbf{1}$ the 3×3 identity tensor. Now, if $\mathbf{s}$ is represented in terms of linear invariants, the following relation holds:

$$\mathbf{t} = \begin{bmatrix} \mathbf{L} & \mathbf{0}_{33} \\ \mathbf{0}_{34} & \mathbf{1} \end{bmatrix} \dot{\mathbf{s}} \tag{3.3.22b}$$

in which $\mathbf{L}$ was defined in eq.(3.2.45*b*). Finally, if Euler parameters are used to define the screw $\mathbf{s}$, one has

$$\mathbf{t} = \begin{bmatrix} \mathbf{E} & \mathbf{0}_{33} \\ \mathbf{0}_{34} & \mathbf{1} \end{bmatrix} \dot{\mathbf{s}} \tag{3.3.22c}$$

where $\mathbf{E}$ is as defined in eq.(3.2.55*b*). Furthermore, the inverse relations of eqs.(3.3.22*a*–*c*) can be readily derived if eqs.(3.2.41*a*), (3.2.44*a*), and (3.2.54*a*) are recalled. Thus, one has

$$\dot{\mathbf{s}} = \begin{bmatrix} \mathbf{B} & \mathbf{0}_{43} \\ \mathbf{0}_{33} & \mathbf{1} \end{bmatrix} \mathbf{t} \tag{3.3.23a}$$

$$\dot{\mathbf{s}} = \begin{bmatrix} \mathbf{\Lambda} & \mathbf{0}_{43} \\ \mathbf{0}_{33} & \mathbf{1} \end{bmatrix} \mathbf{t} \tag{3.3.23b}$$

$$\dot{\mathbf{s}} = \begin{bmatrix} (1/4)\mathbf{E}^T & \mathbf{0}_{43} \\ \mathbf{0}_{33} & \mathbf{1} \end{bmatrix} \mathbf{t} \tag{3.3.23c}$$

depending on whether the natural, the linear, or the quadratic invariants are used to represent both the screw and the twist. Additional results on the general motion of rigid bodies are derived in the next section.

3.4 Further Results on the Velocity Distribution Throughout a Rigid Body

The results that follow were taken from Angeles (1982). They are related to the velocity distribution throughout a rigid body undergoing an arbitrary motion and were first introduced in order to devise a method of computing the angular velocity of a rigid body from the position and velocity of three noncollinear points of the body. Since then, a simplified method was derived, appearing in Angeles (1986-2), that does not require the use of the following theorems. They are included here, nevertheless, for completeness and because of their theoretical merits.

Theorem 3.4.1: *The difference vector of the velocities of any two points of a rigid body undergoing an arbitrary motion is perpendicular to the axis of the instantaneous screw.*

Proof: Let $\mathbf{v}_A$ and $\mathbf{v}_B$ be the velocities of two points, $\mathcal{A}$ and $\mathcal{B}$, of a rigid body. From eq.(3.3.3*a*), one can write the velocity of $\mathcal{B}$ in terms of that of $\mathcal{A}$ as follows:

$$\mathbf{v}_B = \mathbf{v}_A + \mathbf{\Omega}(\mathbf{b} - \mathbf{a}) \tag{3.4.1}$$

Hence, the velocity difference, $\mathbf{d}$, is given by

$$\mathbf{d} \equiv \mathbf{\Omega}(\mathbf{b} - \mathbf{a}) \tag{3.4.2}$$

which makes clear that $\mathbf{d}$ lies in the range of $\mathbf{\Omega}$, and, hence, is perpendicular to ω, which spans its nullspace. Furthermore, ω is parallel to the ISA of the motion, and hence $\mathbf{d}$ is perpendicular to the said axis, thereby proving the theorem.

Theorem 3.4.2: *If the velocities of three noncollinear points of a rigid body are identical, the body undergoes a pure translation.*

Proof: Let $\mathbf{v}_A$, $\mathbf{v}_B$, and $\mathbf{v}_C$ be the respective velocities of points $\mathcal{A}$, $\mathcal{B}$, and $\mathcal{C}$. Now, from eq.(3.3.3*a*), $\mathbf{v}_A$ and $\mathbf{v}_B$ can be written in terms of the velocity and position of $\mathcal{C}$, namely, as

$$\mathbf{v}_A = \mathbf{v}_C + \mathbf{\Omega}(\mathbf{a} - \mathbf{c}) \tag{3.4.3a}$$

$$\mathbf{v}_B = \mathbf{v}_C + \mathbf{\Omega}(\mathbf{b} - \mathbf{c}) \tag{3.4.3b}$$

Since the three velocities are identical, the foregoing equations imply that

$$\mathbf{\Omega}(\mathbf{a} - \mathbf{c}) = \mathbf{0}, \quad \mathbf{\Omega}(\mathbf{b} - \mathbf{c}) = \mathbf{0} \tag{3.4.4}$$

Furthermore, eqs.(3.4.4) imply that both $\mathbf{a} - \mathbf{c}$ and $\mathbf{b} - \mathbf{c}$ lie in the nullspace of $\mathbf{\Omega}$. Since this is one-dimensional, the foregoing implies that $\mathbf{a} - \mathbf{c}$

and $\mathbf{b} - \mathbf{c}$ are parallel to each other, thereby contradicting the hypothesis that the three given points are noncollinear. Thus, the left-hand sides of eqs.(3.4.4) vanish only if $\mathbf{\Omega} = \mathbf{0}$, which means that the body instantaneously undergoes a pure translation, thereby completing the intended proof.

Theorem 3.4.3: *The nonidentical velocities of three points of a rigid body are coplanar if, and only if, one of the following conditions is met:*
i) the motion is a pure rotation;
ii) the motion is general, but the points are collinear; and
iii) the motion is general and the points are noncollinear, but lie in a plane parallel to the ISA.

Proof: First, sufficiency is proved for each of the foregoing cases.

i) If the motion is a pure rotation, it can be assumed without loss of generality that this takes place about the origin, and hence, the velocity of any point of position vector $\mathbf{p}$ can be written as

$$\mathbf{v} = \mathbf{\Omega}\mathbf{p} \tag{3.4.5}$$

which states that $\mathbf{v}$ lies in the range of $\mathbf{\Omega}$, a subspace of dimension 2. This means that the velocity vectors of all the points of the body lie in a plane perpendicular to the nullspace of $\mathbf{\Omega}$, i.e., they are perpendicular to the axis of rotation, thereby showing that the given velocities are coplanar.

ii) Let $\mathcal{A}$, $\mathcal{B}$, and $\mathcal{C}$ be three collinear points of a rigid body, $\mathbf{a}$, $\mathbf{b}$, and $\mathbf{c}$ being their respective position vectors. The velocites of $\mathcal{A}$ and $\mathcal{B}$ can be written in terms of that of $\mathcal{C}$ in the form given by eq.(3.3.3a), namely, as

$$\mathbf{v}_A = \mathbf{v}_C + \mathbf{\Omega}(\mathbf{a} - \mathbf{c}), \quad \mathbf{v}_B = \mathbf{v}_C + \mathbf{\Omega}(\mathbf{b} - \mathbf{c}) \tag{3.4.6}$$

From the assumed collinearity of the three given points, one can write

$$\mathbf{b} - \mathbf{c} = \alpha(\mathbf{a} - \mathbf{c}) \tag{3.4.7}$$

where α is a suitable scalar. Now, from eqs.(3.4.6) and (3.4.7) it follows immediately that

$$\mathbf{v}_B = \alpha \mathbf{v}_A + (1 - \alpha \mathbf{v}_C)$$

thereby proving the coplanarity of the three given velocities.

iii) From eq.(3.3.3a), one can write $\mathbf{v}_A$ and $\mathbf{v}_B$ in terms of $\mathbf{v}_C$, using ω, rather than $\mathbf{\Omega}$, in the form:

$$\mathbf{v}_A = \mathbf{v}_C + \omega \times (\mathbf{a} - \mathbf{c}), \quad \mathbf{v}_B = \mathbf{v}_C + \omega \times (\mathbf{b} - \mathbf{c}) \tag{3.4.8}$$

Now, from the assumption that the three given points lie in a plane parallel to the ISA, ω is obviously parallel to that plane, and hence, two unique scalars, α and β exist, for which one can write

$$\omega = \alpha(\mathbf{a} - \mathbf{c}) + \beta(\mathbf{b} - \mathbf{c}) \tag{3.4.9}$$

Substitution of eq.(3.4.9) into eq.(3.4.8) leads to

$$\mathbf{v}_A = \mathbf{v}_C - \beta\mathbf{p}, \quad \mathbf{v}_B = \mathbf{v}_C + \alpha\mathbf{p} \tag{3.4.10a}$$

where $\mathbf{p}$ is defined as

$$\mathbf{p} = (\mathbf{a} - \mathbf{c}) \times (\mathbf{b} - \mathbf{c}) \tag{3.4.10b}$$

Next, both sides of the first of eqs.(3.4.10a) are multiplied by α, and both sides of the second of those equations are multiplied by β. The addition of the two resulting equations leads to

$$\alpha\mathbf{v}_A + \beta\mathbf{v}_B - (\alpha + \beta)\mathbf{v}_C = \mathbf{0} \tag{3.4.11}$$

thereby making apparent the coplanarity of the three given velocities.

Now, necessity is proved. To this end, the velocities of $\mathcal{A}$ and $\mathcal{B}$ are assumed to be written as in eq.(3.4.8). Next, the three given velocities are assumed to be coplanar, which can be readily expressed as:

$$\mathbf{v}_A \times \mathbf{v}_B \cdot \mathbf{v}_C = 0 \tag{3.4.12}$$

Substitution of eqs.(3.4.8) into eq.(3.4.12) yields, after simplifications,

$$\omega \cdot \mathbf{v}_C[\omega \times (\mathbf{a} - \mathbf{c}) \cdot (\mathbf{b} - \mathbf{c})] = 0 \tag{3.4.13}$$

The left-hand side of eq.(3.4.12) vanishes if one of its two scalar factors does. The vanishing of the first factor implies that there is at least one point of the body, namely, $\mathcal{C}$, whose velocity is perpendicular to the ISA. This means that the sliding of the motion, defined in eq.(3.3.12), vanishes, and hence the body undergoes a pure rotation. Now, the second factor of the left-hand side of eq.(3.4.12) vanishes under either of the two following conditions: i) $(\mathbf{a} - \mathbf{c}) \times (\mathbf{b} - \mathbf{c})$ vanishes, which means that the three given points are collinear, or ii) $\omega \times (\mathbf{a} - \mathbf{c}) \cdot (\mathbf{b} - \mathbf{c})$ vanishes, which means that the three involved vectors are collinear, and hence, the three given points lie in a plane parallel to ω, i.e., to the ISA, thereby completing the intended proof. As direct consequences of the foregoing theorems, one can readily prove the following:

Corollary 3.4.1: *The difference vectors* $\mathbf{v}_A - \mathbf{v}_C$ *and* $\mathbf{v}_B - \mathbf{v}_C$ *are parallel if, and only if, the three given points lie in a plane parallel to the ISA.*

Corollary 3.4.2: *The velocities of any two points of a rigid body cannot be parallel and different, unless the body undergoes a pure rotation.*

Corollary 3.4.3: *If the velocities of any two points of a rigid body are parallel, then either i) the velocities are identical and belong to points lying on a line parallel to the ISA, or ii) those velocities are different, in which case the motion is a pure rotation, and the points lie on a line intersecting the axis of rotation.*

Corollary 3.4.4: *Given three noncollinear points,* $\mathcal{A}$, $\mathcal{B}$, *and* $\mathcal{C}$, *of a rigid body, of position vectors* $\mathbf{a}$, $\mathbf{b}$, *and* $\mathbf{c}$, *respectively, such that* $\mathbf{v}_C = \mathbf{0}$, *if there exists a scalar* β *so that* $\mathbf{v}_B = \beta\mathbf{v}_A$, *then the body undergoes a pure rotation and its axis passes through* $\mathcal{C}$ *and is parallel to vector* $\mathbf{b} - \mathbf{c} - \beta(\mathbf{a} - \mathbf{c})$.

3.5 Compatibility Equations

From the results of Section 3.4, it is apparent that, given the position vectors $\mathbf{p}_i$ and the velocities $\dot{\mathbf{p}}_i$, for $i = 1, 2, 3$, of three *noncollinear* points of a rigid body, then its angular velocity, ω, can be explicitly derived. Indeed, let $\mathbf{c}$ and $\dot{\mathbf{c}}$ denote the position vector of the centroid of the three foregoing points, and its velocity, i.e.,

$$\mathbf{c} \equiv \frac{1}{3}\sum_1^3 \mathbf{p}_i, \quad \dot{\mathbf{c}} \equiv \frac{1}{3}\sum_1^3 \dot{\mathbf{p}}_i \tag{3.5.1}$$

Now, tensors $\mathbf{P}$ and $\dot{\mathbf{P}}$ are defined as

$$\mathbf{P} \equiv [\mathbf{p}_1 - \mathbf{c},\ \mathbf{p}_2 - \mathbf{c},\ \mathbf{p}_3 - \mathbf{c}], \quad \dot{\mathbf{P}} \equiv [\dot{\mathbf{p}}_1 - \dot{\mathbf{c}},\ \dot{\mathbf{p}}_2 - \dot{\mathbf{c}},\ \dot{\mathbf{p}}_3 - \dot{\mathbf{c}}] \tag{3.5.2}$$

Next, eq.(3.3.3*a*) is written for the three given points, in terms of the position and velocity of the centroid $\mathcal{C}$, as

$$\dot{\mathbf{p}}_i - \dot{\mathbf{c}} = \mathbf{\Omega}(\mathbf{p}_i - \mathbf{c}), \quad i = 1, 2, 3 \tag{3.5.3}$$

In light of definitions (3.5.1) and (3.5.2), eqs.(3.5.3) can be rewritten in tensor form as

$$\dot{\mathbf{P}} = \mathbf{\Omega}\mathbf{P} \tag{3.5.4}$$

from which one would like to be able to solve for $\mathbf{\Omega}$, thereby determining the angular velocity sought. This would require multiplying both sides of eq.(3.5.4) by the inverse of $\mathbf{P}$ from the right. However, under the assumption that the three given points are noncollinear, from definitions (3.5.2) it is clear that tensor $\mathbf{P}$ is not invertible, for its columns are linearly dependent, its rank being 2. Nevertheless, the invertibility of $\mathbf{P}$ is not necessary to compute the desired angular velocity. Indeed, if the vector of both sides of eq.(3.5.4) is taken, the following is obtained:

$$\mathbf{T}\omega = \text{vect}(\dot{\mathbf{P}}), \quad \omega = \text{vect}(\mathbf{\Omega}) \tag{3.5.5a}$$

where Theorem 3.2.2 has been invoked, tensor $\mathbf{T}$ thus being defined as

$$\mathbf{T} \equiv \frac{1}{2}[\mathbf{1}\text{tr}(\mathbf{P}) - \mathbf{P}] \tag{3.5.5b}$$

Thus, the sole condition to solve for ω from eq.(3.5.5a) is that $\mathbf{T}$ be invertible. Before deriving this condition, a few facts are introduced, namely,

Theorem 3.5.1: *Tensors* $\mathbf{P}$ *and* $\dot{\mathbf{P}}$, *defined in eq.(3.5.2), are not frame-invariant.*

Proof:

Let $\mathbf{p}_i'$, $\dot{\mathbf{p}}_i'$, $\mathbf{c}'$, and $\dot{\mathbf{c}}'$ denote the position and velocity vectors of points P_i, for $i = 1, 2, 3$, and of their centroid, in a different *observer*—see Section 1.4 for a definition of this term. Moreover, let $\mathbf{Q}$ denote the rotation tensor relating both observers and $\mathbf{d}$ denote the shift of origins between these observers. Thus,

$$\mathbf{p}_i' = \mathbf{Q}\mathbf{p}_i + \mathbf{d}, \quad i = 1, 2, 3 \tag{3.5.6a}$$

Hence,

$$\mathbf{c}' = \mathbf{Q}\mathbf{c} + \mathbf{d} \tag{3.5.6b}$$

Furthermore, upon differentiation of both sides of eq.(3.5.6a), the following is derived:

$$\dot{\mathbf{p}}_i' = \mathbf{Q}\dot{\mathbf{p}}_i + \mathbf{W}\mathbf{p}_i + \dot{\mathbf{d}}, \quad i = 1, 2, 3 \tag{3.5.7a}$$

and

$$\dot{\mathbf{c}}' = \mathbf{Q}\dot{\mathbf{c}} + \mathbf{W}\mathbf{c} + \dot{\mathbf{d}} \tag{3.5.7b}$$

where

$$\mathbf{W} \equiv \dot{\mathbf{Q}}\mathbf{Q}^T \tag{3.5.7c}$$

denotes the angular-velocity tensor of the second observer with respect to the first observer. Hence,

$$\mathbf{P}' = \mathbf{Q}\mathbf{P}, \quad \dot{\mathbf{P}}' = \mathbf{Q}\dot{\mathbf{P}} + \mathbf{W}\mathbf{P} \tag{3.5.8}$$

from which it is obvious that the said tensors are not frame-invariant, i.e., they do not transform as indicated by eq.(1.4.4) under a change of observer, q.e.d.

Furthermore, as a consequence of the lack of invariance of the aforementioned tensors, the moments of $\mathbf{P}'$ and those of $\mathbf{P}$—see Section 1.4—are different, i.e.,

$$\mathrm{tr}(\mathbf{P}')^k = \mathrm{tr}(\mathbf{QP})^k \neq \mathrm{tr}(\mathbf{P})^k, \quad k = 1,2,3 \tag{3.5.9}$$

One more result is the following:

Lemma 3.5.1: *The rank of tensor* $\mathbf{P}$, *defined in eq.(3.5.2), is in general 2. It becomes 1 if, and only if, the three given points are collinear.*

The proof of Lemma 3.5.1 follows from the fact that two column vectors of $\mathbf{P}$ are linearly independent if the three given points are noncollinear. If, on the contrary, these points are collinear, then only one column vector of $\mathbf{P}$ is linearly independent and, vice versa, if $\mathbf{P}$ has only one linearly independent column vector, then the three given points are collinear.

Now, if the three given points are noncollinear, then $\mathbf{P}$ has only one vanishing proper value, the remaining ones being denoted by π_1 and π_2. Let τ_1, τ_2, and τ_3 be the proper values of $\mathbf{T}$. Clearly, the relationship between both sets of proper values if the following:

$$\tau_i = \frac{1}{2}[\mathrm{tr}(\mathbf{P}) - \pi_i], \quad i = 1,2 \tag{3.5.10a}$$

$$\tau_3 = \frac{1}{2}\mathrm{tr}(\mathbf{P}) \tag{3.5.10b}$$

However, $\mathrm{tr}(\mathbf{P})$ is given by

$$\mathrm{tr}(\mathbf{P}) = \pi_1 + \pi_2 \tag{3.5.11a}$$

Upon substitution of eq.(3.5.11a) into eqs.(3.5.10a & b), the following is obtained:

$$\tau_i = \frac{1}{2}(\pi_1 + \pi_2 - \pi_i), \quad i = 1,2; \quad \tau_3 = \frac{1}{2}(\pi_1 + \pi_2) \tag{3.5.11b}$$

Thus, one of the proper values of $\mathbf{T}$ vanishes, and hence, $\mathbf{T}$ is singular, if, and only if, either *i*) $\mathrm{tr}(\mathbf{P})$ is a proper value of $\mathbf{P}$, which means that $\mathbf{P}$ has two vanishing proper values, and hence, $\mathbf{P}$ is of rank 1, or *ii*) $\mathrm{tr}(\mathbf{P})$ vanishes. Condition *i*) means that the three given points are collinear and hence, one has the following:

Theorem 3.5.2: $\mathrm{tr}(\mathbf{P})$, *for* $\mathbf{P}$ *given as in eq.(3.5.2), is a proper value of* $\mathbf{P}$ *if, and only if, the three given points are collinear.*

Moreover,

Corollary 3.5.1: $\text{tr}^2(\mathbf{P}) = \text{tr}(\mathbf{P}^2)$ *if, and only if, the three given points are collinear.*

The proof of Corollary 3.5.1 is straightforward: If the three given points are collinear, then

$$\pi_1 = \text{tr}(\mathbf{P}), \quad \pi_2 = \pi_3 = 0$$

Hence,

$$\text{tr}(\mathbf{P}^2) = \pi_1^2 = \text{tr}^2(\mathbf{P})$$

Similarly, if

$$\text{tr}^2(\mathbf{P}) = \text{tr}(\mathbf{P}^2)$$

then one has

$$\pi_1^2 + 2\pi_1\pi_2 + \pi_2^2 = \pi_1^2 + \pi_2^2$$

which readily leads to

$$\pi_1\pi_2 = 0$$

and hence, one of π_1 and π_2 vanishes, which thus renders $\mathbf{P}$ of rank 1 and, hence, the three given points are collinear.

Now, condition *ii*), namely, $\text{tr}(\mathbf{P}) = 0$, can hold even if the three given points are noncollinear. Indeed, given that $\mathbf{P}$ is not frame-invariant and, particularly, under relations (3.5.9), the value of $\text{tr}(\mathbf{P})$ depends upon the choice of coordinate frame. In fact, for any given set of points $\{P_i\}_1^3$, and a given coordinate frame, a certain coordinate-frame rotation $\mathbf{Q}'$ exists, under which $\text{tr}(\mathbf{P}') = \text{tr}(\mathbf{Q}'\mathbf{P})$ vanishes. In order to show this fact, let $\mathbf{Q}'$ be a rotation about an axis parallel to the nullspace of $\mathbf{P}$, defined by unit vector $\mathbf{e}$, through an angle θ. Moreover, let the Cartesian decomposition of $\mathbf{P}$—see Section 1.2—be given by

$$\mathbf{P} = \mathbf{H} + \mathbf{S}, \quad \mathbf{H} \equiv \frac{1}{2}(\mathbf{P} + \mathbf{P}^T), \quad \mathbf{S} \equiv \frac{1}{2}(\mathbf{P} - \mathbf{P}^T) \tag{3.5.12}$$

Thus,

$$\text{tr}(\mathbf{Q}'\mathbf{P}) = \text{tr}(\mathbf{Q}'\mathbf{H}) + \text{tr}(\mathbf{Q}'\mathbf{S}) \tag{3.5.13}$$

Furthermore, let the Cartesian decomposition of $\mathbf{Q}'$ be given by

$$\mathbf{Q}' = \mathbf{U} + \mathbf{V}, \quad \mathbf{U} \equiv \frac{1}{2}(\mathbf{Q}' + \mathbf{Q}'^T), \quad \mathbf{V} = \frac{1}{2}(\mathbf{Q}' - \mathbf{Q}'^T) \tag{3.5.14a}$$

Thus,

$$\text{tr}(\mathbf{Q}'\mathbf{H}) = \text{tr}(\mathbf{U}\mathbf{H}) + \text{tr}(\mathbf{V}\mathbf{H}) \tag{3.5.14b}$$

where the second term of the right-hand side vanishes because $\mathbf{V}$ is skew-symmetric and $\mathbf{H}$ is symmetric—see Section 1.2. Therefore,

$$\text{tr}(\mathbf{Q}'\mathbf{H}) = \text{tr}(\mathbf{U}\mathbf{H}) = \text{tr}(\mathbf{H}\mathbf{U}) \tag{3.5.14c}$$

Moreover, from eq.(2.4.3c), $\mathbf{U}$ is given by

$$\mathbf{U} = \mathbf{e} \otimes \mathbf{e}(1 - \cos\theta) + \mathbf{1}\cos\theta \tag{3.5.15}$$

Thus,

$$\mathbf{HU} = \mathbf{He} \otimes \mathbf{e}(1 - \cos\theta) + \mathbf{H}\cos\theta$$

Hence,

$$\mathrm{tr}(\mathbf{HU}) = \mathbf{e} \cdot \mathbf{He}(1 - \cos\theta) + \mathrm{tr}(\mathbf{H})\cos\theta$$

However, from eq.(3.5.12),

$$\mathbf{e} \cdot \mathbf{He} = \frac{1}{2}(\mathbf{e}^T\mathbf{Pe} + \mathbf{e}^T\mathbf{P}^T\mathbf{e}) \tag{3.5.16}$$

Since $\mathbf{e}$ lies in the nullspace of $\mathbf{P}$, one has

$$\mathbf{Pe} = \mathbf{0}, \quad \mathbf{e}^T\mathbf{P}^T = \mathbf{0}^T$$

and hence,

$$\mathrm{tr}(\mathbf{HU}) = \mathrm{tr}(\mathbf{H})\cos\theta$$

But

$$\cos\theta = \frac{1}{2}[\mathrm{tr}(\mathbf{Q}') - 1], \quad \mathrm{tr}(\mathbf{H}) = \mathrm{tr}(\mathbf{P}) \tag{3.5.17}$$

Thus,

$$\mathrm{tr}(\mathbf{HU}) = \frac{1}{2}\mathrm{tr}(\mathbf{P})[\mathrm{tr}(\mathbf{Q}') - 1] \tag{3.5.18}$$

thereby completing the computation of the first term of the right-hand side of eq.(3.5.13). On the other hand, the second term of the same expression is readily computed by recalling Theorem 3.2.4, namely,

$$\mathrm{tr}(\mathbf{Q}'\mathbf{S}) = -2\mathrm{vect}(\mathbf{S}) \cdot \mathrm{vect}(\mathbf{Q}') \tag{3.5.19}$$

However,

$$\mathrm{vect}(\mathbf{S}) = \mathrm{vect}(\mathbf{P})$$

and hence,

$$\mathrm{tr}(\mathbf{Q}'\mathbf{P}) = -2\mathrm{vect}(\mathbf{P}) \cdot \mathrm{vect}(\mathbf{Q}') \tag{3.5.20}$$

Finally, eq.(3.5.13) can be written as

$$\mathrm{tr}(\mathbf{Q}')\mathbf{P}) = \frac{1}{2}\mathrm{tr}(\mathbf{P})[\mathrm{tr}(\mathbf{Q}') - 1] - 2\mathrm{vect}(\mathbf{P}) \cdot \mathrm{vect}(\mathbf{Q}') \tag{3.5.21a}$$

which can be rewritten as

$$\mathrm{tr}(\mathbf{Q}'\mathbf{P}) = \frac{1}{2}\mathrm{tr}(\mathbf{P})\cos\theta - 2\mathrm{vect}(\mathbf{P}) \cdot \mathbf{e}\sin\theta \tag{3.5.21b}$$

From eq.(3.5.21*b*) it is clear now that the vanishing of $\mathrm{tr}(\mathbf{Q}'\mathbf{P})$ depends only upon angle θ. If the coefficient of $\sin\theta$ does not vanish in that equation, one can write

$$\theta = \tan^{-1}[\frac{\mathrm{tr}(\mathbf{P})}{4\mathrm{vect}(\mathbf{P})\cdot\mathbf{e}}], \quad \mathrm{vect}(\mathbf{P})\cdot\mathbf{e} \neq 0 \tag{3.5.22a}$$

from which two values of θ can be computed. If that coefficient vanishes, then

$$\theta = \pm\frac{\pi}{2}, \quad \mathrm{vect}(\mathbf{P})\cdot\mathbf{e} = 0 \tag{3.5.22b}$$

thereby proving the following:

Theorem 3.5.3: *A coordinate frame exists for which the trace of* **P** *vanishes, whether the three given points are collinear or not.*

Next, an expression for $\mathbf{T}^{-1}$ is derived, under the assumptions that neither $\mathrm{tr}(\mathbf{P})$ nor $\mathrm{tr}^2(\mathbf{P}) - \mathrm{tr}(\mathbf{P}^2)$ vanishes. From Cayley-Hamilton's Theorem and eq.(3.5.5*b*), $\mathbf{T}^{-1}$ can be expressed as

$$\mathbf{T}^{-1} = \alpha\mathbf{1} + \beta\mathbf{P} + \gamma\mathbf{P}^2 \tag{3.5.23}$$

where coefficients α, β, and γ are to be determined. This can be done by performing the product $\mathbf{T}\mathbf{T}^{-1}$ and equating it with **1**. The said product is obtained as:

$$\mathbf{T}\mathbf{T}^{-1} = \frac{1}{2}\{\alpha\mathrm{tr}(\mathbf{P})\mathbf{1} + [\beta\mathrm{tr}(\mathbf{P}) - \alpha]\mathbf{P} + [\gamma\mathrm{tr}(\mathbf{P}) - \beta]\mathbf{P}^2 - \gamma\mathbf{P}^3\} \tag{3.5.24}$$

Next, $\mathbf{P}^3$ is written as a linear combination of **P** and $\mathbf{P}^3$ by again invoking Cayley-Hamilton's Theorem. In fact, since $\det(\mathbf{P})$ vanishes, the characteristic equation of **P** is the following:

$$\lambda^3 - \mathrm{tr}(\mathbf{P})\lambda^2 + \frac{1}{2}[\mathrm{tr}^2(\mathbf{P}) - \mathrm{tr}(\mathbf{P}^2)]\lambda = 0 \tag{3.5.25a}$$

By virtue of Cayley-Hamilton's Theorem, then, **P** verifies the foregoing equation, i.e.,

$$\mathbf{P}^3 - \mathrm{tr}(\mathbf{P})\mathbf{P}^2 + \frac{1}{2}[\mathrm{tr}^2(\mathbf{P}) - \mathrm{tr}(\mathbf{P}^2)]\mathbf{P} = \mathbf{0} \tag{3.5.25b}$$

Solving for $\mathbf{P}^3$ from eq.(3.5.25*b*), substituting the resulting expression into eq.(3.5.24) and, subsequently, equating the arising expression with **1** yields the following:

$$\frac{\alpha}{2}\mathrm{tr}(\mathbf{P})\mathbf{1} + \frac{1}{2}[\beta\mathrm{tr}(\mathbf{P}) - \alpha + \frac{1}{2}\gamma\mathrm{tr}^2(\mathbf{P}) - \frac{1}{2}\gamma\mathrm{tr}(\mathbf{P}^2)]\mathbf{P} - \frac{1}{2}\beta\mathbf{P}^2 = \mathbf{1} \tag{3.5.26}$$

Now, the coefficients of equal powers of $\mathbf{P}$, appearing on both sides of eq.(3.5.26), are equated, which leads to the following values for α, β, and γ:

$$\alpha = \frac{2}{\mathrm{tr}(\mathbf{P})}, \quad \beta = 0, \quad \gamma = \frac{4}{\mathrm{tr}(\mathbf{P})[\mathrm{tr}(\mathbf{P}^2) - \mathrm{tr}^2(\mathbf{P})]} \tag{3.5.27}$$

One thereby obtains the following expression for $\mathbf{T}^{-1}$:

$$\mathbf{T}^{-1} = \frac{2}{\mathrm{tr}(\mathbf{P})}\mathbf{1} + \frac{4}{\mathrm{tr}(\mathbf{P})[\mathrm{tr}(\mathbf{P}^2) - \mathrm{tr}^2(\mathbf{P})]}\mathbf{P}^2 \tag{3.5.28}$$

Thus, $\mathbf{T}$ is invertible under the conditions assumed at the outset.

Details pertaining to the numerics behind the computation of ω from position and velocity data of three noncollinear points of a rigid body are given in Angeles (1986-2).

Concerning the data of the problem at hand, namely the sets $\{\mathbf{p}_i\}_1^3$ and $\{\dot{\mathbf{p}}_i\}_1^3$, it is apparent that they cannot attain arbitrary values. Indeed, the rigidity conditions that follow should be verified: Let $\mathbf{q}_i$, for $i = 1, 2, 3$, be defined as $\mathbf{q}_i \equiv \mathbf{p}_i - \mathbf{c}$, where $\mathbf{c}$ was defined in eq.(3.5.1). Thus, $\mathbf{q}_i$ is a vector directed from point C, the centroid of the three given points, to point P_i, both of which belong to the same rigid body. Thus, the Euclidean norm of $\mathbf{q}_i$ is constant in time, and hence, its time derivative vanishes, which implies that

$$\mathbf{q}_i^T\dot{\mathbf{q}}_i = 0, \quad i = 1, 2, 3 \tag{3.5.29a}$$

and, since the angle between lines P_iC and P_jC also remains constant in time,

$$\mathbf{q}_i^T\dot{\mathbf{q}}_j = -\dot{\mathbf{q}}_i^T\mathbf{q}_j^T; \quad i, j = 1, 2, 3, \ j \neq i \tag{3.5.29b}$$

which are the compatibility equations sought. Note that the foregoing equations can be written in tensor form as

$$\mathbf{P}^T\dot{\mathbf{P}} = -\dot{\mathbf{P}}^T\mathbf{P} \tag{3.5.30}$$

Clearly, eq.(3.5.30) is merely a direct consequence of eqs.(2.7.6). It is pointed out that eq.(3.5.30) states that tensor $\mathbf{P}^T\dot{\mathbf{P}}$ is skew symmetric if, and only if, the motion of the three points is compatible with the assumption that these are points of the same rigid body.

As discussed in Section 3.4, the three differences of velocities of three noncollinear points of a rigid body lie in a plane perpendicular to ω, which is an alternate way of verifying the compatibility condition. For instance, if the vertices of a regular tetrahedron are labelled A, B, C, and D, and the velocity of A is along edge AD, whereas that of B is along edge BD, then the velocity of C cannot lie on edge CD. It lies in the plane of face ABD.

Chapter 4 ACCELERATION ANALYSIS OF RIGID-BODY MOTIONS

4.1 Introduction

The concept of *angular acceleration* of a rigid-body motion is introduced in this chapter. Moreover, the relationships between the angular acceleration of the rigid body and the time derivatives of the associated rotational invariants are derived, while the acceleration field of the body is derived in terms of the *angular-acceleration tensor.* Unlike the displacement and the velocity fields, the acceleration field, in general, contains one point of zero acceleration that is unique. Indeed, as shown in this chapter, the angular-acceleration tensor is, in general, nonsingular, and becomes singular only in the particular case in which the body *instantaneously* undergoes a rotation about a stationary axis. Thus, in the case of acceleration fields, no theorem similar to that of Chasles' for displacement fields, or its counterpart for velocity fields, exists.

4.2 Acceleration Field of a Body Moving About a Fixed Point

The velocity field of a rigid body moving about a fixed point, which is assumed to be the origin of $\mathcal{E}^3$, with angular velocity $\mathbf{\Omega}$, is given by eq.(3.2.3). Next this is reproduced for quick reference, the position vector of the moving point in the current configuration being denoted by $\mathbf{p}$. Thus,

$$\mathbf{v} = \mathbf{\Omega}\mathbf{p} \tag{4.2.1}$$

Further, both sides of eq.(4.2.1) are differentiated with respect to time, which yields

$$\dot{\mathbf{v}} = \dot{\mathbf{\Omega}}\mathbf{p} + \mathbf{\Omega}\dot{\mathbf{p}} \tag{4.2.2}$$

where $\dot{\mathbf{p}}$ is readily identified as $\mathbf{v}$ of eq.(4.2.1). Substitution of this equation into eq.(4.2.2) yields, in turn,

$$\dot{\mathbf{v}} = (\dot{\mathbf{\Omega}} + \mathbf{\Omega}^2)\mathbf{p} \tag{4.2.3a}$$

Now, the sum in parentheses in eq.(4.2.3) is defined as the *angular-acceleration tensor*, $\mathbf{\Psi}$, i.e.,

$$\mathbf{\Psi} \equiv \dot{\mathbf{\Omega}} + \mathbf{\Omega}^2 \tag{4.2.3b}$$

and hence, $\dot{\mathbf{v}}$ can be written as:

$$\dot{\mathbf{v}} = \mathbf{\Psi}\mathbf{p} \tag{4.2.3c}$$

It is apparent that $\dot{\mathbf{\Omega}}$ is skew symmetric, whereas $\mathbf{\Omega}^2$ is symmetric. Hence. eq.(4.2.3*b*) represents the *Cartesian decomposition* of $\mathbf{\Psi}$. The *angular-acceleration vector*, $\dot{\omega}$, is now defined as the vector of $\mathbf{\Psi}$, i.e.,

$$\dot{\omega} \equiv \text{vect}(\mathbf{\Psi}) \tag{4.2.4}$$

which clearly equals the vector of $\dot{\mathbf{\Omega}}$ alone.

Moreover, $\dot{\mathbf{\Omega}}$ and $\mathbf{\Omega}^2$ are orthogonal, for

$$\text{tr}(\dot{\mathbf{\Omega}}\mathbf{\Omega}^2) = 0 \tag{4.2.5}$$

which is a direct consequence of $\dot{\mathbf{\Omega}}$ being skew symmetric, and $\mathbf{\Omega}^2$ being symmetric.

It is apparent from the foregoing that $\mathbf{\Psi}$ is neither symmetric nor skew symmetric, and may or may not be singular. The conditions under which $\mathbf{\Psi}$ is nonsingular are derived next.

Clearly, $\mathbf{\Psi}$ is singular if, and only if, the nullspaces of $\mathbf{\Omega}$ and $\dot{\mathbf{\Omega}}$ have a common subspace—other than $\mathbf{0}$, of course. Now, let $\mathbf{f}$ be a unit vector in the nullspace of $\mathbf{\Omega}$, i.e.,

$$\mathbf{\Omega}\mathbf{f} = \mathbf{0} \tag{4.2.6}$$

The time derivative of $\mathbf{f}$, $\dot{\mathbf{f}}$, thus verifies

$$\mathbf{\Omega}\dot{\mathbf{f}} = -\dot{\mathbf{\Omega}}\mathbf{f} \tag{4.2.7a}$$

as well as

$$\mathbf{f}^T\dot{\mathbf{f}} = 0 \tag{4.2.7b}$$

From eq.(4.2.7*a*) it is apparent that $\mathbf{f}$ is in the nullspace of $\dot{\mathbf{\Omega}}$ if, and only if, $\dot{\mathbf{f}}$ either vanishes or lies in the nullspace of $\mathbf{\Omega}$. In the first case, the motion is one in which the body under study rotates instantaneously about a fixed axis. In the second case, $\dot{\mathbf{f}}$ is parallel to ω. It is then shown that $\dot{\mathbf{f}}$, if not zero, is perpendicular to ω, and hence lies in the range of $\mathbf{\Omega}$, not in its nullspace. To this end, eqs.(4.2.7a & b) are solved for $\dot{\mathbf{f}}$. These equations are first written as:

$$\mathbf{M}\dot{\mathbf{f}} = \mathbf{b} \tag{4.2.8a}$$

with the following definitions:

$$\mathbf{M} = \begin{bmatrix} \mathbf{\Omega} \\ \mathbf{f}^T \end{bmatrix}, \quad \mathbf{b} = \begin{bmatrix} -\dot{\mathbf{\Omega}}\mathbf{f} \\ 0 \end{bmatrix} \tag{4.2.8b}$$

Thus, $\dot{\mathbf{f}}$ can be obtained from eq.(4.2.8b) in terms of the Moore-Penrose generalized inverse of $\mathbf{M}$, which yields

$$\dot{\mathbf{f}} = \mathbf{M}^I\mathbf{b} \tag{4.2.9a}$$

and

$$\mathbf{M}^I = (\mathbf{M}^T\mathbf{M})^{-1}\mathbf{M}^T \tag{4.2.9b}$$

provided that $\mathbf{M}^T\mathbf{M}$ is invertible. But this is so, because

$$\mathbf{M}^T\mathbf{M} = -\mathbf{\Omega}^2 + \mathbf{f} \otimes \mathbf{f} \tag{4.2.10}$$

from which it is not obvious that $\mathbf{M}^T\mathbf{M}$ is nonsingular. This becomes apparent once the proper values of the aforementioned tensor are found, and none of these is shown to vanish identically. Clearly, $\mathbf{f}$ is a proper vector of $\mathbf{M}^T\mathbf{M}$, of proper value $+1$, which can be readily verified by making use of eq.(4.2.6). If now $\mathbf{g}$ is defined as a unit vector perpendicular to $\mathbf{f}$, one has, from eq.(4.2.10),

$$\mathbf{M}^T\mathbf{M}\mathbf{g} = -\mathbf{\Omega}^2\mathbf{g} \tag{4.2.11}$$

and hence, any proper vector of $\mathbf{\Omega}^2$ is also a proper vector of $\mathbf{M}^T\mathbf{M}$. Moreover, the proper values of $\mathbf{M}^T\mathbf{M}$, apart from $+1$, are the negatives of the nonzero proper values of $\mathbf{\Omega}^2$. However, from Theorem 3.2.1, it is apparent that the nonzero proper values of $\mathbf{\Omega}^2$ are both $\mathrm{tr}(\mathbf{\Omega}^2)/2$. Thus, the proper values of $\mathbf{M}^T\mathbf{M}$ are the following:

$$\lambda_1 = 1, \quad \lambda_2 = \lambda_3 = -\frac{1}{2}\mathrm{tr}(\mathbf{\Omega}^2)$$

none of which vanishes if $\mathbf{\Omega}$ itself does not vanish. Hence, $\mathbf{M}^T\mathbf{M}$ is invertible. Having shown that the aforementioned tensor is invertible, one now can proceed to compute its inverse. This is most simply done by assuming that the said inverse has the following form:

$$(\mathbf{M}^T\mathbf{M})^{-1} = \alpha\mathbf{\Omega}^2 + \beta\mathbf{f} \otimes \mathbf{f} \tag{4.2.12}$$

which is plausible in view of eq.(4.2.10). Scalars α and β are, as yet, undefined. These are determined next by computing the following product:

$$(\mathbf{M}^T\mathbf{M})^{-1}\mathbf{M}^T\mathbf{M} = -\alpha\mathbf{\Omega}^4 + \beta\mathbf{f} \otimes \mathbf{f} \tag{4.2.13a}$$

By virtue of eq.(3.2.14), $\mathbf{\Omega}^4$ can be written as

$$\mathbf{\Omega}^4 = \frac{1}{2}\mathrm{tr}(\mathbf{\Omega}^2)\mathbf{\Omega}^2 \tag{4.2.13b}$$

and hence,

$$(\mathbf{M}^T\mathbf{M})^{-1}\mathbf{M}^T\mathbf{M} = -\frac{\alpha}{2}\mathrm{tr}(\mathbf{\Omega}^2)\mathbf{\Omega}^2 + \beta\mathbf{f}\otimes\mathbf{f} \tag{4.2.13c}$$

Application of eq.(1.2.9) produces

$$\mathbf{\Omega}^2 = -\mathbf{1} + \omega\otimes\omega \tag{4.2.14a}$$

On the other hand, $\mathbf{f}$ can be written as

$$\mathbf{f} = \frac{\omega}{\|\omega\|} \tag{4.2.14b}$$

from which it is apparent that the tensor product $\mathbf{f}\otimes\mathbf{f}$ can be written as

$$\mathbf{f}\otimes\mathbf{f} = \frac{\omega\otimes\omega}{\|\omega\|^2} \tag{4.2.14c}$$

Moreover, by virtue of eq.(3.2.21), one has

$$\|\omega\|^2 = -\frac{\mathrm{tr}(\mathbf{\Omega}^2)}{2} \tag{4.2.14d}$$

Substitution of eq.(4.2.14d) into eq.(4.2.14d) produces

$$\mathbf{f}\otimes\mathbf{f} = -2\omega\otimes\omega/\mathrm{tr}(\mathbf{\Omega}^2) \tag{4.2.14e}$$

Furthermore, upon substitution of eqs.(4.2.14a & e) into eq.(4.2.13c), one obtains

$$(\mathbf{M}^T\mathbf{M})^{-1}\mathbf{M}^T\mathbf{M} = \frac{\alpha}{2}\mathrm{tr}(\mathbf{\Omega}^2)\mathbf{1} - [\frac{\alpha}{2}\mathrm{tr}(\mathbf{\Omega}^2) + \frac{2\beta}{\mathrm{tr}(\mathbf{\Omega}^2)}]\omega\otimes\omega \tag{4.2.14f}$$

Now α and β are determined by setting the coefficients of $\mathbf{1}$ and $\omega\otimes\omega$ in eq.(4.2.14f) equal to 1 and 0, respectively, which produces

$$\alpha = 2/\mathrm{tr}(\mathbf{\Omega}^2), \quad \beta = -1/\alpha \tag{4.2.15}$$

Upon substitution of eq.(4.2.15) into eq.(4.2.12), the following expression for $(\mathbf{M}^T\mathbf{M})^{-1}$ is obtained:

$$(\mathbf{M}^T\mathbf{M})^{-1} = \frac{2}{\mathrm{tr}(\mathbf{\Omega}^2)}\mathbf{1} - \frac{\mathrm{tr}(\mathbf{\Omega}^2)}{2}\mathbf{f}\otimes\mathbf{f} \tag{4.2.16}$$

and hence,

$$\mathbf{M}^I = \frac{1}{2\mathrm{tr}(\mathbf{\Omega}^2)}[\,-4\mathbf{\Omega}^2,\ [4 - \mathrm{tr}(\mathbf{\Omega}^2)\,]\mathbf{f}\,] \tag{4.2.17}$$

Substitution of eq.(4.2.17) into eq.(4.2.9a) yields

$$\dot{\mathbf{f}} = \frac{2}{\mathrm{tr}(\mathbf{\Omega}^2)}\mathbf{\Omega}^2\dot{\mathbf{\Omega}}\mathbf{f} \tag{4.2.18a}$$

which can be written alternatively as

$$\dot{\mathbf{f}} = -\frac{1}{\|\omega\|^2}\omega \times [\,\omega \times (\dot{\omega} \times \mathbf{f})\,] \tag{4.2.18b}$$

thereby obtaining an expression for the time rate of change of the unit vector defining the direction of the instantaneous axis of rotation. From eq.(4.2.18a), it is apparent that if $\dot{\mathbf{f}}$ does not vanish, then it lies in the range of $\mathbf{\Omega}$; eq.(4.2.18b) shows, in turn, that if $\dot{\mathbf{f}}$ does not vanish, then it is orthogonal to ω. One then has proven the following:

Theorem 4.2.1: *The angular-velocity tensor and its time rate of change have a common nullspace if, and only if, the body undergoes instantaneously a rotation about a fixed axis.*

Moreover,

Corollary 4.2.1: *The angular-acceleration tensor is invertible, unless the body undergoes instantaneously a rotation about a fixed axis.*

Similar results were derived by Veldkamp (1969), by resorting to a coordinate-dependent formulation.

Next, the relationships between $\dot{\omega}$ and the time derivatives of the natural, the linear, and the quadratic invariants of the rotation involved, are derived. The first relationships are readily obtained from eqs.(3.2.41a & b). Differentiation of both sides of this equation with respect to time yields

$$\ddot{\nu} = \mathbf{B}\dot{\omega} + \mathbf{B}'\omega \tag{4.2.19a}$$

where

$$\ddot{\nu} = \begin{bmatrix} \ddot{\mathbf{e}} \\ \ddot{\theta} \end{bmatrix}, \quad \mathbf{B}' = \begin{bmatrix} \mathbf{U} \\ \dot{\mathbf{e}}^T \end{bmatrix} \tag{4.2.19b}$$

where $\mathbf{U}$ is given, in turn, as

$$\mathbf{U} = \frac{\mathbf{U}'}{D} \tag{4.2.19c}$$

with the following definitions:

$$\begin{aligned} \mathbf{U}' = & -2\mathbf{e}\cdot\omega\mathbf{1} + 2(3-\cos\theta)\mathbf{e}\cdot\omega\mathbf{e}\otimes\mathbf{e} - 2\omega\otimes\mathbf{e} \\ & - 2\cos\theta\mathbf{e}\otimes\omega - \sin\theta(\omega\otimes\mathbf{e} + \mathbf{e}\otimes\omega) \\ & - \sin\theta\mathbf{1}\times(\omega - \mathbf{e}\cdot\omega\mathbf{e}) \end{aligned} \tag{3.2.19d}$$

$$D = 4(1-\cos\theta) \tag{3.2.19e}$$

The inverse relation of eq.(3.2.19a) can most readily be derived by differentiating with respect to time both sides of eq.(3.2.43c), which yields

$$\dot{\omega} = \mathbf{N}\ddot{\nu} + \dot{\mathbf{N}}\dot{\nu} \tag{4.2.20a}$$

where $\dot{\mathbf{N}}$ is given, in turn, as

$$\dot{\mathbf{N}} = [\, \dot{\theta}\cos\theta\mathbf{1} + \dot{\theta}\sin\theta\mathbf{1} \times \mathbf{e},\ \dot{\mathbf{e}} \,] \tag{4.2.20b}$$

Now the relationships between $\dot{\omega}$ and the time derivatives of the linear invariants of the rotation under study are derived. To this end, both sides of eq.(3.2.44a) are differentiated with respect to time, thus obtaining

$$\ddot{\lambda} = \mathbf{\Lambda}\dot{\omega} + \dot{\mathbf{\Lambda}}\omega \tag{4.2.21a}$$

where

$$\ddot{\lambda} = \begin{bmatrix} \ddot{\mathbf{q}} \\ \ddot{q}_0 \end{bmatrix}, \quad \dot{\mathbf{\Lambda}} = \begin{bmatrix} \frac{1}{2}[\mathrm{tr}(\dot{\mathbf{R}})\mathbf{1} - \dot{\mathbf{R}}] \\ -\dot{\mathbf{q}}^T \end{bmatrix} \tag{4.2.21b}$$

Now, from eq.(3.2.33),

$$\mathrm{tr}(\dot{\mathbf{R}}) = -2\dot{\theta}\sin\theta$$

and, from eq.(3.2.32),

$$\mathrm{tr}(\dot{\mathbf{R}}) = -2\omega \cdot \mathbf{e}\sin\theta = -2\omega \cdot \mathbf{q} \tag{4.2.21c}$$

Moreover, from eq.(3.2.2b),

$$\dot{\mathbf{R}} = \mathbf{\Omega}\mathbf{R} = \omega \times \mathbf{R} \tag{4.2.21d}$$

and, from eqs.(3.2.44a & b),

$$\dot{\mathbf{q}} = \frac{1}{2}[\mathrm{tr}(\mathbf{R})\mathbf{1} - \mathbf{R}]\omega \tag{4.2.21e}$$

Hence, $\dot{\mathbf{\Lambda}}$ can be written as

$$\dot{\mathbf{\Lambda}} = \begin{bmatrix} -(\omega \cdot \mathbf{q})\mathbf{1} - \frac{1}{2}\omega \times \mathbf{R} \\ -\frac{1}{2}\omega^T[\mathbf{1}\mathrm{tr}(\mathbf{R}) - \mathbf{R}^T] \end{bmatrix} \tag{4.2.22}$$

The inverse relationships, i.e., those producing $\dot{\omega}$ in terms of $\ddot{\lambda}$, can be most readily derived by differentiation of both sides of eq.(3.2.45a) with respect to time, thus obtaining

$$\dot{\omega} = \mathbf{L}\ddot{\lambda} + \dot{\mathbf{L}}\dot{\lambda} \tag{4.2.23a}$$

where

$$\dot{\mathbf{L}} = -\frac{1}{1+q_0}[\, \dot{\mathbf{q}} \otimes \mathbf{q} + \mathbf{q} \otimes \dot{\mathbf{q}} - \mathbf{1} \times \dot{\mathbf{q}} - \frac{\mathbf{q} \otimes \mathbf{q} - \mathbf{1} \times \mathbf{q}}{1+q_0}\dot{q}_0,\ (1+q_0)\dot{\mathbf{q}} \,] \tag{4.2.23b}$$

Upon expansion of the right-hand side of eq.(4.2.23a), one obtains, for $q_0 \neq -1$,

$$\dot{\omega} = \ddot{\mathbf{q}} + \frac{\mathbf{1} \times \mathbf{q} - \mathbf{q} \otimes \mathbf{q}}{1 + q_0}\ddot{\mathbf{q}} - \mathbf{q}\ddot{q}_0 + \mathbf{C}(\mathbf{q}, q_0, \dot{\mathbf{q}}, \dot{q}_0)\dot{\mathbf{q}} \qquad (4.2.23c)$$

where tensor $\mathbf{C}(\mathbf{q}, q_0, \dot{\mathbf{q}}, \dot{q}_0)$ is defined as

$$\begin{aligned}\mathbf{C}(\mathbf{q}, q_0, \dot{\mathbf{q}}, \dot{q}_0) = & \frac{\dot{q}_0}{(1 + q_0)^2}[\mathbf{q} \otimes \mathbf{q} - \mathbf{1} \times \mathbf{q} - (1 + q_0)^2\mathbf{1}] \\ & - \frac{1}{1 + q_0}[\mathbf{q} \otimes \dot{\mathbf{q}} + \mathbf{q} \cdot \dot{\mathbf{q}}\mathbf{1}] \end{aligned} \qquad (4.2.23d)$$

For $q_0 = -1$, i.e., for $\theta = \pi$, $\dot{\omega}$ can be obtained either by formally taking the limit of the right-hand side of eq.(4.2.23c), when $\theta \to \pi$, or by simply substituting $\theta = \pi$ into eq.(4.2.20b). In either case, one obtains

$$\dot{\omega} = 2\dot{\mathbf{e}} \times \ddot{\mathbf{e}} + \mathbf{e}\ddot{\theta}, \quad \text{for} \quad \theta = \pi \qquad (4.2.24)$$

However, eq.(4.2.24) does not contain the linear invariants and their time derivatives. The corresponding explicit relationship can be obtained, nevertheless, by introducing eq.(3.2.52) and similar ones derived upon differentiation of eqs.(3.2.47a & b), when evaluated at $\theta = \pi$. Time differentiation of the said equations produces

$$\begin{aligned}\mathbf{q}^{(3)} = & \, \mathbf{e}^{(3)} \sin\theta + 2\ddot{\mathbf{e}}\dot{\theta}\cos\theta + 3\dot{\mathbf{e}}\ddot{\theta}\cos\theta - 3\dot{\mathbf{e}}\dot{\theta}^2\sin\theta \\ & - 3\dot{\mathbf{e}}\dot{\theta}\ddot{\theta}\sin\theta - \mathbf{e}\dot{\theta}^3\cos\theta \end{aligned} \qquad (4.2.25a)$$

$$q_0^{(3)} = -3\dot{\theta}\ddot{\theta}\cos\theta + \dot{\theta}^3\sin\theta - \theta^{(3)}\sin\theta \qquad (4.2.25b)$$

where $(\cdot)^{(3)}$ stands for the third time derivative of $(\cdot)$. Hence,

$$\mathbf{q}^{(3)} = -2\ddot{\mathbf{e}}\dot{\theta} - 3\dot{\mathbf{e}}\ddot{\theta} + \mathbf{e}\dot{\theta}^3, \quad q_0^{(3)} = 3\dot{\theta}\ddot{\theta}, \quad \text{for} \quad \theta = \pi \qquad (4.2.26)$$

Substitution of the foregoing equations into eq.(4.2.12) yields

$$\dot{\omega} = \frac{\dot{\mathbf{q}}}{\ddot{q}_0} \times (\mathbf{q}^{(3)} - \frac{q_0^{(3)}}{2\ddot{q}_0}\ddot{\mathbf{q}}) - \frac{q_0^{(3)}\dot{\mathbf{q}}}{3\ddot{q}_0}, \quad \text{for} \quad \theta = \pi \qquad (4.2.27)$$

Now the relationships between $\dot{\omega}$ and the Euler parameters, and their time derivatives, are derived. Differentiation of both sides of eq.(3.2.55a) with respect to time yields

$$\dot{\omega} = \mathbf{E}\ddot{\eta} + \dot{\mathbf{E}}\dot{\eta} \qquad (4.2.28a)$$

However, as one can readily verify,

$$\dot{\mathbf{E}}\dot{\eta} = \mathbf{0} \qquad (4.2.28b)$$

and hence,

$$\dot{\omega} = \mathbf{E}\ddot{\eta} \tag{4.2.28c}$$

The inverse relation of eq.(4.2.28c) can be readily derived as the minimum-norm solution of that equation. This is obtained in terms of the generalized inverse $\mathbf{E}^{\dagger}$. Recalling expression (3.2.55c) for the said inverse, then, one has

$$\ddot{\eta} = \frac{1}{4}\mathbf{E}^T\dot{\omega} \tag{4.2.29}$$

which is in agreement with a previous result by Nikravesh, Wehage and Kwon (1985).

4.3 Acceleration Field of a Rigid Body Under General Motion

Given in eq.(3.3.3a) is the velocity $\dot{\mathbf{p}}$ of a point $\mathcal{P}$ of position vector $\mathbf{p}$, of a rigid body undergoing a general motion. That equation relates the said velocity with that of a point $\mathcal{A}$, of position vector $\mathbf{a}$, and velocity $\dot{\mathbf{a}}$, of the same rigid body. Differentiation of both sides of the aforementioned equation with respect to time yields

$$\ddot{\mathbf{p}} = \ddot{\mathbf{a}} + \dot{\mathbf{\Omega}}(\mathbf{p} - \mathbf{a}) + \mathbf{\Omega}(\dot{\mathbf{p}} - \dot{\mathbf{a}}) \tag{4.3.1a}$$

Substitution of $\dot{\mathbf{p}} - \dot{\mathbf{a}}$ from eq.(3.3.3a) into eq.(4.3.1a) yields

$$\ddot{\mathbf{p}} = \ddot{\mathbf{a}} + (\dot{\mathbf{\Omega}} + \mathbf{\Omega}^2)(\mathbf{p} - \mathbf{a}) \tag{4.3.1b}$$

or, in terms of the angular-acceleration tensor,

$$\ddot{\mathbf{p}} = \ddot{\mathbf{a}} + \mathbf{\Psi}(\mathbf{p} - \mathbf{a}) \tag{4.3.1c}$$

Thus, the acceleration field of a rigid body undergoing a general motion is given by eq.(4.3.1c) in terms of the known position and acceleration of a *base point* $\mathcal{A}$, and the angular-acceleration tensor of the motion under study.

From the discussion of Section 4.2, $\mathbf{\Psi}$ is, in general, nonsingular, and hence, a point $\mathcal{P}_0$ can be found, with position vector $\mathbf{p}_0$, whose acceleration vanishes. This position vector is readily computed from eq.(4.3.1c) by setting $\ddot{\mathbf{p}}$ equal to zero. This yields

$$\mathbf{p}_0 = \mathbf{a} - \mathbf{\Psi}^{-1}\ddot{\mathbf{a}} \tag{4.3.2}$$

and $\mathcal{P}_0$ is referred to as *the acceleration pole* of the rigid-body motion —see, e.g., Veldkamp (1969).

Now, if the body under study rotates instantaneously about a stationary axis, then $\mathbf{\Psi}$ is singular, and no single acceleration pole exists. Rather, a set of zero-acceleration points exists that lies on that axis.

4.4 Change of Observer. Coriolis' Theorem

This section is devoted to one of the most useful results in the development of kinematics, namely, *Coriolis' Theorem*. Within the spirit of this book, namely, that of invariance, this theorem would not be needed. However, for the sake of completeness, and in order to clarify a few misconceptions in this regard, this result is discussed here. Since this discussion departs from the unified approach adopted at the outset, a caveat is in order, namely, vector and tensor components will be needed here, and hence, the following discussion refers to component transformation, rather than to physical relations as such.

In dealing with systems of rigid bodies moving with respect to each other, the motion description in a single reference frame, or observer, may be either impractical or unfeasible. This description is generally eased if referred to intermediate moving frames. For example, let $\mathcal{A}$ and $\mathcal{B}$ be two reference frames, also called *coordinate systems* or observers, that move with respect to each other. The relative motion of these is equivalent to that of rigid bodies, and hence, $\mathcal{A}$ and $\mathcal{B}$ can be regarded equivalently as such. However, a clear distinction must now be made, when speaking of *components* of vectors and higher-order tensors, as to the frame of reference. Thus, if the origins of $\mathcal{A}$ and $\mathcal{B}$ coincide, and $\mathbf{Q}$ is the rotation carrying $\mathcal{A}$ into $\mathcal{B}$, the components of the position vector $\mathbf{p}$ of a point $\mathcal{P}$ will be indicated as $[\mathbf{p}]_{\mathcal{A}}$ or $[\mathbf{p}]_{\mathcal{B}}$, depending on which reference frame is being used. That is, $[\mathbf{p}]_{\mathcal{X}}$ is an array of real numbers representing the coordinates of $\mathbf{p}$ in frame $\mathcal{X}$. Clearly, the coordinates of $\mathcal{P}$ in $\mathcal{A}$ are identical to those of $\mathcal{P}'$ in $\mathcal{B}$, where $\mathcal{P}'$ denotes the position vector of $\mathcal{P}$ when regarded as a point of $\mathcal{A}$ after this frame has undergone rotation $\mathbf{Q}$. Hence, the position vector of $\mathcal{P}'$ is $\mathbf{Qp}$, and one has

$$[\mathbf{p}]_{\mathcal{A}} = [\mathbf{Qp}]_{\mathcal{B}} \tag{4.4.1a}$$

Clearly, eq.(4.4.1a) can be rewritten as

$$[\mathbf{p}]_{\mathcal{A}} = [\mathbf{Q}]_{\mathcal{B}}[\mathbf{p}]_{\mathcal{B}} \tag{4.4.1b}$$

and hence

$$[\mathbf{p}]_{\mathcal{B}} = [\mathbf{R}]_{\mathcal{B}}[\mathbf{p}]_{\mathcal{A}} \tag{4.4.2}$$

where, clearly, $\mathbf{R} = \mathbf{Q}^T$ is the rotation carrying $\mathcal{B}$ into $\mathcal{A}$. In the following discussion, $\mathcal{B}$ is arbitrarily regarded as a *fixed frame* $\mathcal{F}$, and $\mathcal{A}$ as a *moving frame* $\mathcal{M}$, $\mathbf{R}$ being defined, then, as

$$\mathbf{R}\colon \mathcal{F} \rightarrow \mathcal{M} \tag{4.4.3a}$$

and the following general relation holds:

$$[\mathbf{p}]_{\mathcal{F}} = [\mathbf{R}]_{\mathcal{F}}[\mathbf{p}]_{\mathcal{M}} \tag{4.4.3b}$$

If, however, the origins of $\mathcal{F}$ and $\mathcal{M}$ do not coincide, let $\mathbf{r}$ be the vector joining them, directed from the former to the latter. Then, clearly, eq.(4.4.3b) transforms into

$$[\mathbf{p}]_{\mathcal{F}} = [\mathbf{r}]_{\mathcal{F}} + [\mathbf{R}]_{\mathcal{F}}[\mathbf{p}]_{\mathcal{M}} \tag{4.4.4}$$

Differentiation of both sides of eq.(4.4.4) with respect to time yields

$$[\dot{\mathbf{p}}]_{\mathcal{F}} = [\dot{\mathbf{r}}]_{\mathcal{F}} + [\dot{\mathbf{R}}]_{\mathcal{F}}[\mathbf{p}]_{\mathcal{M}} + [\mathbf{R}]_{\mathcal{F}}[\dot{\mathbf{p}}]_{\mathcal{M}} \tag{4.4.5a}$$

Now the following identity is introduced in eq.(4.4.5a):

$$\dot{\mathbf{R}} \equiv \dot{\mathbf{R}}\mathbf{R}^T\mathbf{R} \tag{4.4.5b}$$

which allows one to rewrite the said equation as

$$[\dot{\mathbf{p}}]_{\mathcal{F}} = [\dot{\mathbf{r}}]_{\mathcal{F}} + [\dot{\mathbf{R}}\mathbf{R}^T]_{\mathcal{F}}[\mathbf{R}]_{\mathcal{F}}[\mathbf{p}]_{\mathcal{M}} + [\mathbf{R}]_{\mathcal{F}}[\dot{\mathbf{p}}]_{\mathcal{M}} \tag{4.4.5c}$$

Since $\dot{\mathbf{R}}\mathbf{R}^T$ is the angular-velocity tensor, $\mathbf{\Omega}$, of the motion of $\mathcal{M}$ with respect to $\mathcal{F}$, eq.(4.4.5c) readily leads to

$$[\dot{\mathbf{p}}]_{\mathcal{F}} = [\dot{\mathbf{r}}]_{\mathcal{F}} + [\mathbf{\Omega}]_{\mathcal{F}}[\mathbf{R}]_{\mathcal{F}}[\mathbf{p}]_{\mathcal{M}} + [\mathbf{R}]_{\mathcal{F}}[\dot{\mathbf{p}}]_{\mathcal{M}} \tag{4.4.6}$$

Upon differentiation of both sides of eq.(4.4.6) with respect to time, one obtains

$$\begin{aligned}[\ddot{\mathbf{p}}]_{\mathcal{F}} =& [\ddot{\mathbf{r}}]_{\mathcal{F}} + [\dot{\mathbf{\Omega}}]_{\mathcal{F}}[\mathbf{R}]_{\mathcal{F}}[\mathbf{p}]_{\mathcal{M}} + [\mathbf{\Omega}]_{\mathcal{F}}[\dot{\mathbf{R}}]_{\mathcal{F}}[\mathbf{p}]_{\mathcal{M}} + \\ & [\mathbf{\Omega}]_{\mathcal{F}}[\mathbf{R}]_{\mathcal{F}}[\dot{\mathbf{p}}]_{\mathcal{M}} + [\dot{\mathbf{R}}]_{\mathcal{F}}[\dot{\mathbf{p}}]_{\mathcal{M}} + [\mathbf{R}]_{\mathcal{F}}[\ddot{\mathbf{p}}]_{\mathcal{M}}\end{aligned} \tag{4.4.7a}$$

By taking into account eq.(4.4.5b), the third and the fifth terms of the right-hand side of eq.(4.4.7a) can be rewritten as

$$[\mathbf{\Omega}]_{\mathcal{F}}[\dot{\mathbf{R}}]_{\mathcal{F}}[\mathbf{p}]_{\mathcal{M}} = [\mathbf{\Omega}^2]_{\mathcal{F}}[\mathbf{R}]_{\mathcal{F}}[\mathbf{p}]_{\mathcal{M}} \tag{4.4.7b}$$

$$[\dot{\mathbf{R}}]_{\mathcal{F}}[\dot{\mathbf{p}}]_{\mathcal{M}} = [\mathbf{\Omega}]_{\mathcal{F}}[\mathbf{R}]_{\mathcal{F}}[\dot{\mathbf{p}}]_{\mathcal{M}} \tag{4.4.7c}$$

Substitution of eqs.(4.4.7b & c) into eq.(4.4.7a) yields

$$[\ddot{\mathbf{p}}]_{\mathcal{F}} = [\ddot{\mathbf{r}}]_{\mathcal{F}} + [\mathbf{\Psi}]_{\mathcal{F}}[\mathbf{R}]_{\mathcal{F}}[\mathbf{p}]_{\mathcal{M}} + 2[\mathbf{\Omega}]_{\mathcal{F}}[\mathbf{R}]_{\mathcal{F}}[\dot{\mathbf{p}}]_{\mathcal{M}} + [\mathbf{R}]_{\mathcal{F}}[\ddot{\mathbf{r}}]_{\mathcal{M}} \tag{4.4.8}$$

where definition (4.2.3b) has been recalled. Furthermore, the third term of the right-hand side of eq.(4.4.8) is the so-called *Coriolis acceleration*, after the French mathematician to whom it is ascribed (Coriolis 1835). This term, however, had been anticipated by Euler (1750). That equation constitutes, then, *Coriolis' Theorem*.

From the foregoing it is clear that the Coriolis-acceleration term arises from the description adopted, namely, via moving observers, and hence, contrary to a popular belief, *it bears no physical significance.*

4.5 Determination of Angular Acceleration. Compatibility Equations

Two questions are answered in this section. The first one is: Given the acceleration field in a rigid body undergoing an arbitrary motion, what is its angular acceleration? The second one is: Given an acceleration field, is it associated with the motion undergone by a rigid body? In the foregoing, the acceleration field is assumed to be given via the motion of three noncollinear points of the body, and not as a function of the position vector and time, throughout the whole body. Were this the case, then a simple test would suffice. Indeed, assume that the motion is given as follows: Let $\mathbf{p}'$ be the position vector of a point of the body undergoing that motion, in its current configuration, $\mathbf{p}$ being the corresponding position vector in a reference configuration. The *deformation gradient*, $\mathbf{F}$, is defined as (Truesdell 1960):

$$\mathbf{F} \equiv \frac{\partial \mathbf{p}'}{\partial \mathbf{p}}$$

From eq.(2.5.4) it is clear that the motion is rigid if, and only if, $\mathbf{F}$ is both a function of time only, and proper orthogonal. In dealing with the velocity field, assume that this is given as $\dot{\mathbf{p}} = \dot{\mathbf{p}}(\mathbf{p}, t)$, where $\dot{\mathbf{p}}$ is the velocity of a given point of position vector $\mathbf{p}$, at time t. The *velocity gradient* of this motion, $\mathbf{L}$, is given as

$$\mathbf{L} \equiv \frac{\partial \dot{\mathbf{p}}}{\partial \mathbf{p}}$$

From eq.(3.3.3a), it is apparent that the motion is rigid if, and only if, $\mathbf{L}$ is, first of all, a function of time t only, and skew symmetric, thereby being identical to the angular-velocity tensor, $\mathbf{\Omega}(t)$. Furthermore, if the acceleration field is given as $\ddot{\mathbf{p}} = \ddot{\mathbf{p}}(\mathbf{p}, t)$, then the *acceleration gradient*, $\mathbf{A}$, is defined as

$$\mathbf{A} \equiv \frac{\partial \ddot{\mathbf{p}}}{\partial \mathbf{p}}$$

From eq.(4.3.1c), it is clear that the motion is rigid if, and only if, $\mathbf{A}$ is, first of all, a function of time t only, and its symmetric component, $(\mathbf{A} + \mathbf{A}^T)/2$, is identical to $\mathbf{\Omega}^2$. In this case, then, the skew-symmetric component of $\mathbf{A}$, $(\mathbf{A} - \mathbf{A}^T)/2$, is identical to the time derivative of $\mathbf{\Omega}$, i.e., to $\dot{\mathbf{\Omega}}$. The foregoing would then answer the second question. In the context of rigid-body motions, however, most likely the motion is described in terms of those of a finite set of points of the body. Moreover, from Sections 2.7 and 3.5, it is clear that the motions of three noncollinear points of a rigid body determine the motion of the body. Thus, the first question could be rephrased as: Given the positions, velocities, and accelerations of three noncollinear points of a rigid body, what is its angular acceleration? This question is answered next.

If the motions of three noncollinear points of a rigid body are given, then the angular-velocity tensor, $\boldsymbol{\Omega}$, can be readily computed as outlined in Section 3.5. Hence, the symmetric component, $\boldsymbol{\Omega}^2$, of the angular-acceleration tensor, $\boldsymbol{\Psi}$, is already available. Thus, determining $\boldsymbol{\Psi}$ reduces to determining its skew-symmetric component, $\dot{\boldsymbol{\Omega}}$. Now, since this component is fully determined by its vector, $\dot{\omega}$, three independent equations are needed to fully determine the angular acceleration under study. It is shown next that these equations can be derived if the motions of three noncollinear points of the rigid body are known.

Following the notation introduced in Section 3.4, the accelerations of three noncollinear points, P_1, P_2, and P_3, are denoted by $\ddot{\mathbf{p}}_1$, $\ddot{\mathbf{p}}_2$, and $\ddot{\mathbf{p}}_3$, respectively. Furthermore, the acceleration of the centroid of the given points, C, of position vector $\mathbf{c}$, is denoted by $\ddot{\mathbf{c}}$, and is given by

$$\ddot{\mathbf{c}} = \frac{1}{3}\sum_{1}^{3} \ddot{\mathbf{p}}_i \tag{4.5.1}$$

Now, tensor $\ddot{\mathbf{P}}$ is defined as

$$\ddot{\mathbf{P}} = [\,\ddot{\mathbf{p}}_1 - \ddot{\mathbf{c}},\, \ddot{\mathbf{p}}_2 - \ddot{\mathbf{c}},\, \ddot{\mathbf{p}}_3 - \ddot{\mathbf{c}}\,] \tag{4.5.2}$$

From eq.(4.3.1c), the accelerations $\{\ddot{\mathbf{p}}_i\}_1^3$ can be written in terms of $\mathbf{c}$ and $\ddot{\mathbf{c}}$ as

$$\ddot{\mathbf{p}}_i = \ddot{\mathbf{c}} + \boldsymbol{\Psi}(\mathbf{p}_i - \mathbf{c}), \quad i = 1, 2, 3 \tag{4.5.3a}$$

where $\boldsymbol{\Psi}$ is the angular-acceleration tensor defined in eq.(4.2.3b). From eq.(4.5.3a) one readily obtains

$$\ddot{\mathbf{P}} = \boldsymbol{\Psi}\mathbf{P} \tag{4.5.3b}$$

with $\mathbf{P}$ defined as in Section 3.4, i.e., as

$$\mathbf{P} = [\,\mathbf{p}_1 - \mathbf{c},\, \mathbf{p}_2 - \mathbf{c},\, \mathbf{p}_3 - \mathbf{c}\,] \tag{4.5.3c}$$

When introducing the Cartesian decomposition of $\boldsymbol{\Psi}$, eq.(4.2.3b), into eq.(4.5.3b), this equation leads to

$$\dot{\boldsymbol{\Omega}}\mathbf{P} = \ddot{\mathbf{P}} - \boldsymbol{\Omega}^2\mathbf{P} \tag{4.5.4}$$

Now, the vector of both sides of eq.(4.5.4) is taken, and Theorem 3.2.2 is applied to the left-hand side of the resulting equation. This yields

$$\mathbf{T}\dot{\omega} = \mathrm{vect}(\ddot{\mathbf{P}} - \boldsymbol{\Omega}^2\mathbf{P}) \tag{4.5.5}$$

where $\mathbf{T}$ was defined in eq.(3.4.5b), and is reproduced next for quick reference

$$\mathbf{T} = \frac{1}{2}[\,\mathbf{1}\mathrm{tr}(\mathbf{P}) - \mathbf{P}\,] \tag{4.5.6}$$

From the foregoing and the results of Section 3.5 it follows readily that $\dot{\omega}$ can be determined from eq.(4.5.5) if, and only if, the given points are noncollinear and tr($\mathbf{P}$) does not vanish, which thus renders $\mathbf{T}$ invertible, and hence,

$$\dot{\omega} = \mathbf{T}^{-1}\mathrm{vect}(\ddot{\mathbf{P}} - \mathbf{\Omega}^2\mathbf{P}) \tag{4.5.7}$$

Now it is clear that the position vectors, the velocities, and the accelerations of three noncollinear points of a rigid body determine the acceleration field throughout the body. Moreover, from eq.(4.5.7) it is apparent that the velocities of the three points themselves do not appear in the expression for $\dot{\omega}$. What appears therein, instead, is the angular velocity and the position vectors and accelerations of the three given points of the body. The three accelerations, however, cannot be given arbitrarily, for they must conform to a compatibility equation, which is next derived.

Differentiation of both sides of eq.(3.4.9) with respect to time yields

$$\mathbf{P}^T\ddot{\mathbf{P}} + \ddot{\mathbf{P}}^T\mathbf{P} + 2\dot{\mathbf{P}}^T\dot{\mathbf{P}} = \mathbf{0} \tag{4.5.8}$$

a result that could have been arrived at from eq.(4.5.4), as well. It is interesting to note that tensor $\mathbf{P}^T\ddot{\mathbf{P}} + \ddot{\mathbf{P}}^T\mathbf{P}$, appearing in eq.(4.5.8), is negative definite and, in fact, equals $-2\dot{\mathbf{P}}^T\dot{\mathbf{P}}$. Moreover, from eq.(4.5.8) it follows that

$$\mathrm{tr}(\mathbf{P}^T\ddot{\mathbf{P}}) = -\mathrm{tr}(\dot{\mathbf{P}}^T\dot{\mathbf{P}})$$

thereby completing the study of rigid-body motions. The numerics involved in the computation of the angular-acceleration tensor of a rigid body motion, from the positions, velocities, and accelerations of three noncollinear points of the body, is given in Angeles (1987).

Chapter 5 KINEMATIC CHAINS

5.1 Introduction

The coupling of rigid bodies by means of mechanical constraints constitutes a kinematic chain. This coupling takes place pairwise, and hence, it is given the name *kinematic pair.* In this chapter the basic classification of kinematic pairs, namely, *lower* and *upper* kinematic pairs, is introduced and the discussion will be mainly devoted to a study of the former. The latter are discussed briefly in Section 5.8. Furthermore, kinematic chains coupled by lower kinematic pairs are classified into *simple* and *complex.* The former, in turn, can be either *open* or *closed.* In any case, the degree of freedom of the chain is determined resorting either to a *Chebyshev-Grübler-Kutzbach formula* or to the *Jacobian matrix* of the chain under study. It is shown that the said type of formulae, based solely on the topology of the chain, has limited applicability regarding the determination of the chain's degree of freedom. On the other hand, the Jacobian of the chain provides a widely applicable means of determining the degree of freedom of not only simple, but also complex kinematic chains. Regarding the latter, two particular types of kinematic structures are distinguished, namely, tree structures and chains with multiple closed loops. The former are discussed briefly, for they are not essential in this context; the latter are studied in detail regarding the determination of their degree of freedom. Next, an item that is of the utmost relevance in dynamics is introduced, namely, the kinematic constraint equations of a general mechanical system. This leads to the concept of independent and dependent *generalized speeds*, by means of which the *twists* of all the links of the chain are expressed as linear transformations of the vector of independent generalized speeds. Finally, the problems of analysis and synthesis of kinematic chains are outlined briefly in Sections 5.9 and 5.10, respectively, but methods of solution of these problems, not being within the scope of this book, are not considered herein.

5.2 Kinematic Pairs. Classification

A *kinematic pair* is the coupling of *two rigid bodies*, which thus produces a *constrained* relative motion of one body with respect to the other. Basically two different types of kinematic pairs exist, namely, *lower* and *upper* kinematic pairs. A lower kinematic pair exists when the coupling between the two bodies takes place along a surface; an upper kinematic pair exists when the coupling takes place along a line or a point. The basic lower kinematic pairs are the following: *i*) *rotational* , *ii*) *prismatic*, *iii*) *screw*, *iv*) *cylindrical*, *v*) *spherical*, and *vi*) *planar*. The foregoing pairs are usually denoted by the following letters: R for rotational, P for prismatic, H for screw, C for cylindrical, S for spherical, and E for planar. In the rotational pair, contact takes place along a circular cylindrical surface, in such a way that the two coupled bodies can undergo a relative rotation about the axis of the cylinder, but any sliding is prevented. In the prismatic pair, contact takes place along a prismatic surface, and hence, the two bodies can undergo a translation along the edges of the prism. In the screw pair, contact takes place along a surface resembling the thread of a screw of pitch p, and hence, the bodies undergo a relative rotation about and a relative translation along the axis of the screw. Furthermore, if the relative rotation is denoted by θ and the relative translation by u, then u and θ are related by $u = p\theta$. Moreover, when two bodies are coupled by a cylindrical pair, contact takes place along a circular cylindrical surface, such as in the case of the rotational pair; however, in the cylindrical pair the bodies can translate freely along the axis of the cylinder. In the spherical pair, contact takes place along a spherical surface, and hence, the two bodies can rotate about a common point, namely, the center of the sphere. Finally, in a planar pair, contact takes place along a plane, and hence, the bodies can translate along two independent directions parallel to the plane, and rotate about an axis perpendicular to the plane. In fact, the screw, cylindrical, spherical, and planar pairs can be obtained as a composition of the rotational and the prismatic pairs, which are henceforth termed *elementary pairs*. Indeed, the screw and the cylindrical pairs are derived from a rotational and a prismatic pair of parallel axes; the spherical pair is derived, in turn, from three rotational pairs of concurrent axes; and the planar pair is derived from two prismatic pairs of non parallel axes and a rotational pair of axis perpendicular to the plane determined by the two foregoing axes.

A cylinder, whether circular or not, and a sphere rolling on a plane, are examples of bodies coupled by an upper kinematic pair. In the first case, contact takes place along a line, namely, an element of the cylinder; in the second along a point, i.e., the contact point between the sphere and the plane. This chapter focuses on lower kinematic pairs; upper kinematic pairs will be treated in Section 5.8 in the context of general holonomic and nonholonomic constraints.

5.3 Simple Kinematic Chains. Degree of Freedom by Groups of Motion

The practice followed so far, of denoting points with caligraphic fonts, is abandoned in this and the succeeding sections. Points are now denoted with italics, because caligraphic fonts will be reserved for groups of motions associated with kinematic pairs and combinations thereof.

A kinematic chain is said to be *simple* if each of its links is coupled to, at most, two other links. If all its links are coupled to two other links, then the chain is *closed*; otherwise, it is open. Note that, in an open kinematic chain, all links are coupled to other two links, except for the two *end links*, which are coupled to only one other link. Therefore, in a simple kinematic chain, whether closed or open, one can number the links successively from 1 to n *uniquely*, once the first link has been defined, and the direction of ordering is established. Closed kinematic chains constitute *linkages*, whereas open kinematic chains, *manipulators*. In a closed kinematic chain the $(n+1)$st link is identical to the first one. In an open kinematic chain one distinguishes between the *fixed link* and the *end link*, which are the two links of the chain coupled to one single link. The fixed link is usually assigned the number 0, while the end link is usually given the number n. Clearly, a single rigid body can be regarded as a simple kinematic chain; in fact, it is the simplest possible kinematic chain. Moreover, according to the foregoing definitions, a single rigid body is an open kinematic chain.

The *degree of freedom* of a kinematic chain is defined as the necessary and sufficient number of variables that define uniquely the position and orientation of all the links of the chain. Thus, the degree of freedom of a single rigid body is 6. The effect of coupling of rigid bodies produces a loss of degree of freedom. For instance, a rotational pair allows only a relative rotation about one axis, and hence, the coupling reduces by 5 the overall degree of freedom of the two bodies, viewed as a two-link, open kinematic chain. Therefore, the degree of freedom of the relative motion of the two bodies thus coupled is 1. It can also be said that one of the two foregoing bodies moves with respect to the other with a *single* degree of freedom. The same holds for the prismatic and the screw pairs. On the other hand, two bodies coupled by a cylindrical pair move with respect to each other with a *double* degree of freedom, whereas two bodies coupled either by a spherical or a planar pair move with respect to each other with a *triple* degree of freedom.

The relative motion associated with each of the foregoing lower kinematic pairs constitutes a *group* under the operation of motion composition, if the said motion is regarded as a particular type of *affine transformation.* Indeed, if one of the coupled bodies is given two successive motions with respect to the other body, then the resulting single motion is of the same nature as the individual motions. For example, for the R pair, the resulting motion of two successive rotations about the R axis is again a rotation about the same axis, its angle being the sum of the individual angles. A similar relationship holds for the P and the H pairs. In the same fashion, the composition of two individual motions of one body coupled to another by a C pair is another combination of rotation about and translation along the axis of the pair, both the angle of rotation and the resulting displacement being the sum of the individual ones. Furthermore, the composition of motions allowed by an S pair is another rotation about the center of the pair, and the composition of two motions allowed by an E pair is again a planar motion in the plane of E. The set of motions undergone by an unconstrained rigid body also constitutes a group, namely, the *group of rigid-body motions.* Henceforth, the group of rigid-body motions will be denoted by $\mathcal{G}$, and the groups associated with each of the six lower kinematic pairs will be regarded as *subgroups* of $\mathcal{G}$. The notation introduced to denote the different subgroups of interest is described next.

Let the axis of a given R pair pass through a certain point A and have a direction given by that of the unit vector $\mathbf{e}$. The subgroup associated with this pair will be indicated as $\mathcal{R}(A,\mathbf{e})$. Similarly, let the direction of the relative translation provided by a prismatic pair P be given by a unit vector $\mathbf{e}$; then, the subgroup associated with P will be denoted by $\mathcal{P}(\mathbf{e})$. Regarding the screw pair, let its axis pass through point A and have a direction given by the unit vector $\mathbf{e}$, its pitch being p. This pair has a subgroup that will be denoted by $\mathcal{H}(A,p,\mathbf{e})$. Furthermore, let the axis of a given cylindrical pair pass through a point A and have a direction given by the unit vector $\mathbf{e}$, the associated subgroup thus being denoted by $\mathcal{C}(A,\mathbf{e})$. Now, if the center of a spherical pair is denoted by A, the subgroup associated with this pair is denoted by $\mathcal{S}(A)$. Finally, let the relative motions associated with a certain planar pair take place on a plane perpendicular to the unit vector $\mathbf{e}$; then, the subgroup associated with this pair is denoted by $\mathcal{E}(\mathbf{e})$. Notice that the position of the aforementioned plane is irrelevant, and hence, only its normal vector, $\mathbf{e}$, need be specified. The quantities in parentheses associated with the foregoing subgroups can be called the *subgroup parameters.* In the following discussion, whenever the subgroup parameters are either self-understood or irrelevant for the discussion at hand, they will be omitted.

If the set of motions of a subgroup $\mathcal{G}_1$ are contained in subgroup $\mathcal{G}_2$, the relation between both groups will be indicated as

$$\mathcal{G}_1 \subset \mathcal{G}_2$$

Moreover, the *identity* subgroup, denoted by $\mathcal{I}$, is defined as that describing the relative motion between two rigid bodies which are rigidly coupled and hence, this subgroup represents the absence of relative motion. Furthermore, any two subgroups, $\mathcal{G}_1$ and $\mathcal{G}_2$, are said to be *independent* if their intersection is the identity subgroup. From the foregoing discussion, one has the following results:

$$\mathcal{R}(A,\mathbf{e}) \subset \mathcal{C}(A,\mathbf{e})$$
$$\mathcal{P}(\mathbf{e}) \subset \mathcal{C}(A,\mathbf{e})$$
$$\mathcal{H}(A,p,\mathbf{e}) \subset \mathcal{C}(A,\mathbf{e})$$
$$\mathcal{R}(A,\mathbf{e}) \subset \mathcal{S}(A)$$
$$\mathcal{R}(A,\mathbf{e}) \subset \mathcal{E}(\mathbf{e})$$

Furthermore, if A and B are any two points, and segment AB is not parallel to $\mathbf{e}$, then

i) $\mathcal{R}(A,\mathbf{e})$ and $\mathcal{R}(B,\mathbf{e})$ are independent.

ii) $\mathcal{H}(A,p,\mathbf{e})$ and $\mathcal{H}(B,p,\mathbf{e})$ are independent.

iii) $\mathcal{H}(A,p,\mathbf{e})$ and $\mathcal{H}(A,q,\mathbf{e})$ are independent, provided $p \neq q$.

The *dimension* of one of the foregoing subgroups is defined as the degree of freedom of the relative motion it allows (Hervé 1978). It is indicated as dim$(\cdot)$, where $(\cdot)$ denotes any one of those subgroups. Clearly, the dimension of $\mathcal{R}$, $\mathcal{P}$, and $\mathcal{H}$ is 1, that of $\mathcal{C}$ is 2, whereas that of $\mathcal{S}$ and $\mathcal{E}$ is 3. Moreover,

$$\dim(\mathcal{G}) = 6, \quad \dim(\mathcal{I}) = 0$$

Additional subgroups of $\mathcal{G}$, not associated with any lower pair, are next defined:

The relative translation between two rigid bodies along two independent directions is referred to as the *plane translation*. If the two directions are contained in a plane perpendicular to the unit vector $\mathbf{e}$, then, this subgroup is denoted by $\mathcal{T}(\mathbf{e})$. Of course, $\dim(\mathcal{T})=2$.

The relative translation of two rigid bodies along three noncoplanar directions is called the *three-dimensional translation* and is denoted by $\mathcal{U}$, where, obviously, no subgroup parameters need be indicated. Clearly, $\dim(\mathcal{U})=3$.

Two more similar subgroups are next defined. These are the *translating screw* and the *translating gimbal*. The former, $\mathcal{Y}(p,\mathbf{e})$, denotes the subgroup of motions undergone by a screw of pitch p and axis parallel to the unit vector $\mathbf{e}$, the said axis being able to undergo translations in two independent directions perpendicular to $\mathbf{e}$. The latter, $\mathcal{X}(\mathbf{e})$, denotes the subgroup of motions undergone by the rotor of a gimbal whose spin axis is parallel to unit vector $\mathbf{e}$, the gimbal frame being capable of translating freely in space.

With the foregoing background, the concept of kinematic pair can be extended. Assume that the *subchain* composed of links numbered i, $i+1, \ldots, i+j$, of a given kinematic chain, is simple. Hervé (1978) defines the *bond—liaison* in French—between links i and $i+j$ as the *product* of the subgroups $\mathcal{G}_k$, for $k = i, i+1, \ldots, i+j-1$, in which $\mathcal{G}_k$ is the subgroup of relative motions associated with the kinematic pair connecting the kth and the $(k+1)$st links. Clearly, the product of two subgroups is to be interpreted as the composition of the motions they represent, i.e., as a product of transformations. Let the said bond be denoted by $\mathcal{B}_{i,i+j}$. Then,

$$\mathcal{B}_{i,i+j} \equiv \Pi_i^{i+j-1} \mathcal{G}_k \tag{5.3.1}$$

Clearly, the resulting bond need not be a subgroup itself. For example, if $\mathcal{G}_1 = \mathcal{R}(A, \mathbf{e}_1)$ and $\mathcal{G}_2 = \mathcal{R}(B, \mathbf{e}_2)$, then $\mathcal{B}_{1,3}$ is not a group, whether A and B are coincident or not. Indeed, if A and B are coincident, then the resulting motion is a subset of $\mathcal{S}(A)$, but not all of it. Moreover, from the results of Chapter 2, the aforementioned resulting motion is neither a rotation about $\mathbf{e}_1$ nor about $\mathbf{e}_2$. On the other hand, if the said points are not coincident, then the resulting motion is more complex, for it is no longer a subset of the spherical subgroup.

Furthermore, the dimension of a bond is defined in exactly the same manner as that of a subgroup, and, clearly,

$$\dim(\mathcal{B}_{1,3}) = \dim(\mathcal{G}_1) + \dim(\mathcal{G}_2) - \dim(\mathcal{G}_1 \cap \mathcal{G}_2) \tag{5.3.2}$$

where $\mathcal{G}_1$, $\mathcal{G}_2$, and $\mathcal{B}_{1,3}$ are as previously defined.

From the foregoing, then, it is clear that the dimension of a bond composed of $j+1$ links, comprising j kinematic pairs, is, at most, equal to the sum of the dimensions of the individual pairs. Let $m \geq 0$ be the difference between the two aforementioned quantities. That is,

$$m \equiv \sum_i^{i+j-1} \dim(\mathcal{G}_k) - \dim(\mathcal{B}_{i,i+j}) \tag{5.3.3}$$

In an open kinematic chain, for example, with links numbered from 0 to n, where 0 is the *fixed link* and n is the *end link*, the degree of freedom of the end link, f, is simply the dimension of $\mathcal{B}_{1,n}$, i.e.,

$$f = \dim(\mathcal{B}_{1,n}) \tag{5.3.4}$$

In an n-link closed kinematic chain, let m be defined as

$$m \equiv \sum_1^n \dim(\mathcal{G}_k) - \dim(\mathcal{B}_{1,n+1}) \tag{5.3.5}$$

Thus, m represents the *idle* degree of freedom of the chain, i.e., the number of independent single-degree-of-freedom motions that do not affect the motion of the overall chain. This concept will be illustrated with examples, after introducing a few additional definitions.

Once the 0th and the nth links of a simple kinematic chain have been defined, its *topology* is indicated as follows: Let K_i be the kinematic pair coupling the ith and the $(i+1)$st links of the chain. Then, the said topology is given as $K_0 K_1 \cdots K_{n-1}$. If the foregoing chain is closed, then the description $K_0 K_1 \cdots K_n$ makes sense, for the last pair is understood as the one coupling the nth and the $(n+1)$st links of the chain, the latter being identical to the first link. For example, if $RCCC$ denotes a closed kinematic chain, it is composed of 4 pairs, and hence, 4 links, which are coupled by 1 rotational and 3 cylindrical pairs.

From the foregoing discussion, it is apparent that the degree of freedom of an open kinematic chain need not be identical to that of its end link. As a matter of fact, the latter is, at most, equal to the former. In a closed kinematic chain, the two links coupled to the fixed one are usually referred to, respectively, as the input or driving and the output or driven links. Similar to open chains, in a closed chain the degree of freedom of the output link is, at most, equal to that of the chain itself. In summary, one can say that the degree of freedom of any link of a kinematic chain, whether open or closed, cannot be greater than that of the chain itself.

Some examples that illustrate the foregoing ideas are given below. First consider an open SS chain. This is a 3-link chain containing two spherical pairs, the subgroups associated with each of these being denoted by $\mathcal{G}_1$ and $\mathcal{G}_2$. Thus,

$$\dim(\mathcal{G}_1) = \dim(\mathcal{G}_2) = 3$$

Moreover, let A and B denote the centers of each of the two S pairs of the chain under discussion, and let $\mathbf{e}$ denote the unit vector parallel to $\overline{AB}$. Clearly,

$$\mathcal{G}_1 \cap \mathcal{G}_2 = \mathcal{R}(A, \mathbf{e})$$

and hence,

$$\dim(\mathcal{G}_1 \cap \mathcal{G}_2) = 1$$

From eq.(5.3.2), then,

$$\dim \mathcal{B}_{1,3} = 3 + 3 - 1 = 5$$

and hence, the degree of freedom of the end link is 5, whereas that of the chain is 6. The kinematic interpretation of the foregoing result is such that, whereas the end link can attain an arbitrary orientation, its point B is constrained to remain a fixed distance, say d, from point A. Thus, if the screw $\mathbf{s}$, defining the *pose*—position and orientation—of the third link, is given in terms of the linear invariants, grouped in the 4-dimensional vector λ —introduced in Section 2.4—and the position vector $\mathbf{b}$ of B, then, a total of 7 parameters is needed to specify the pose of the end link. These, however, are not independent, for they are constrained by the following:

$$\lambda^T \lambda = 1, \quad \mathbf{b}^T \mathbf{b} = d$$

and hence the degree of freedom of the chain under study is, in fact, 5. Similarly, given an open ES kinematic chain, the degree of freedom of its end link is 5, because the intersection of the groups associated with the E and the S pairs is $\mathcal{R}(A, \mathbf{e})$, where A is the point of the plane of E on which S is centered, and $\mathbf{e}$ is the unit normal of E. Thus, the dimension of the intersection of the subgroups of E and S is 1, and the result immediately follows.

As a further example, consider the $RSSR$ closed kinematic chain. The degree of freedom of its input and output links is 1, but that of the chain is 2. This is due to the fact that, once the input link is fixed, the output link can attain only one of two possible configurations (Lichtenheldt and Luck 1979) that are called *conjugate*, but the intermediate link, also called the *coupler link*, can undergo an arbitrary rotation about the axis defined by the centers of the two S pairs.

When considering the degree of freedom of the end link of an open, simple kinematic chain, or of that of the output link of a closed chain, this is usually referred to as the degree of freedom of the chain itself. Henceforth, this practice will be followed because, in reality, the degree of freedom of one particular link is of relevance, whereas that of the chain itself is seldom of concern, at least in the case of simple kinematic chains.

A basic problem in the study of kinematic chains is determining the degree of freedom of the chain. In this regard, closed kinematic chains are far more challenging than open chains. In fact, the degree of freedom of the end link of an open kinematic chain or, in this context, of the chain itself, is simply defined as the dimension of the associated bond—see eq.(5.3.4)—which can be determined without any major difficulty. A problem arises when dealing with closed kinematic chains, as will become apparent in the following discussion.

For purposes of the problem at hand, however, kinematic chains will be classified into two major types, namely, those whose degree of freedom can be determined from their sole topology and those whose topology does not suffice to determine their degree of freedom. The degree of freedom of the former can be determined with the sole aid of the theory of groups, as noted by Hervé (1978), whereas that of the latter must be determined with methods considering the geometry of the chain, as well. The first type is further divided into two subtypes, namely, *trivial* and *exceptional*.

Trivial kinematic chains are those whose kinematic pairs have associated subgroups whose product is, in turn, a subgroup of the group of rigid-body motions. On the contrary, a kinematic chain whose associated subgroups have a product which is not a subgroup of the group of rigid-body motions, but that can be *reduced* to an *equivalent* trivial chain, is an exceptional chain. Thus, chains whose associated subgroups have a product that is not a subgroup of the rigid-body group, and that cannot be reduced to a trivial chain, belong to the second type, and are termed *paradoxical* or *overconstrained*. First, the degree of freedom of trivial chains is determined; then, the concept of equivalent kinematic chain is introduced, and a method for determining the degree of freedom of exceptional chains is derived. Finally, a few examples of paradoxical kinematic chains are given, the determination of their degree of freedom to be undertaken in the following section.

Let the product of all subgroups of a trivial $(n+1)$-link kinematic chain be a subgroup $\mathcal{G}'$ of $\mathcal{G}$. Moreover, let d be the dimension of $\mathcal{G}'$, i.e., if $\mathcal{G}_i$ denotes the subgroup associated with the kinematic pair coupling the ith and the $(i+1)$st links, then

$$\Pi_0^n \mathcal{G}_i = \mathcal{G}' \subset \mathcal{G}, \quad \dim(\mathcal{G}') = d \tag{5.3.6}$$

Furthermore, let the ith kinematic pair, K_i impose r_i constraints on the two bodies it couples. Then, the degree of freedom of the $n+1$ links before coupling is simply dn, for an arbitrary link, e.g., the 0th, is considered fixed. Hence, the degree of freedom of the chain, after coupling, is given by

$$f = dn - \sum_0^n r_i - m \tag{5.3.7}$$

where m is defined in eq.(5.3.5). Eq.(5.3.7) is a *Chebyshev-Grübler-Kutzbach* formula. For instance, in the case of the $RSSR$ linkage, the product of all associated subgroups is $\mathcal{G}$ itself, and hence, the chain is trivial. Moreover, $d = 6$, $n = 3$, $r_0 = 5$, $r_1 = 3$, $r_2 = 3$, $r_3 = 5$ and $m = 1$. Application of the foregoing formula produces $f = 1$, which is the correct degree of freedom of the chain.

Now, exceptional chains are studied. To this end, the concept of a *reduced equivalent* chain is first introduced. Since there are several steps involved, these steps will be numbered:

i) Decompose all pairs other than R and P into a combination of these. Now, one has two chains, the original one and the decomposed one.

ii) Assume the original chain has $n + 1$ links and the decomposed one $n' + 1$, where, clearly, $n' \geq n$. Now, form the products of all posssible combinations of 2, 3, ..., $n' + 1$ links. Clearly, the number of possible combinations of $k \leq n' + 1$ links, N_k, is

$$N_k = \binom{n'+1}{k} \equiv \frac{(n'+1)!}{(n'+1-k)!k!} \tag{5.3.8}$$

Hence, the total number of all possible combinations, N, is given by

$$N = \sum_{2}^{n'+1} N_k \tag{5.3.9}$$

iii) Out of the arising N possible products, sort all those that are subsets of $\mathcal{G}$. If, except for the two-link products, none is a subgroup of $\mathcal{G}$, the chain is paradoxical and its degree of freedom cannot be determined from its sole topology.

iv) Let $\mathcal{G}_M$ be the subgroup, of those sorted out in *iii*), with the greatest dimension, d_M. If this is not unique, then redefine $\mathcal{G}_M$ as the product of all those different subgroups of dimension d_M. If this product is a subgroup of $\mathcal{G}$, then the chain is exceptional and the procedure can be continued. If it is not, then the chain is paradoxical.

v) Intersect $\mathcal{G}_M$ with each of the $\mathcal{G}_i$ subgroups of the original chain. Clearly, each of these intersections is a subgroup of $\mathcal{G}$. Let $\mathcal{G}'_i$ be the arising intersection.

vi) Let $\mathcal{G}''$ be the product of all $\mathcal{G}'_i$ subgroups. If $\mathcal{G}'' = \mathcal{G}_M$, then the chain is exceptional. Otherwise, it is paradoxical.

vii) If the chain is exceptional, form a new kinematic chain with pairs K'_i defined as indicated below. If $\mathcal{G}'_i$ is a subgroup associated with one of the six lower kinematic pairs, then K'_i is simply defined as that pair. If it is not, then p K'_{ij} pairs, for $j = 1, 2, \ldots, p$, are defined, which yield $\mathcal{G}'_i$ as their product.

viii) The resulting kinematic chain is the reduced equivalent kinematic chain of the original one. It is trivial and hence, its degree of freedom can be obtained with the Chebyshev-Grübler-Kutzbach formula given before. Clearly, both chains have the same degree of freedom.

As an example of the above reduction, consider the familiar *RRRC* kinematic chain appearing in all types of crank-slider mechanisms in internal combustion engines or air compressors. It can be readily realized that this chain is not trivial, for the product of all its associated pair subgroups is not a subgroup of $\mathcal{G}$. Indeed, the associated subgroups are the following: $\mathcal{R}(A, \mathbf{e}_1)$, $\mathcal{R}(B, \mathbf{e}_1)$, $\mathcal{R}(C, \mathbf{e}_1)$, and $\mathcal{C}(C, \mathbf{e}_2)$, where $\mathbf{e}_2$ is a unit vector parallel to the plane defined by A, B, and C, and $\mathbf{e}_1$ is a unit vector perpendicular to this plane. The product of the first three pairs can be readily found to be $\mathcal{E}(\mathbf{e}_1)$. However, the product of this subgroup with $\mathcal{C}(C, \mathbf{e}_2)$ is not a subgroup of $\mathcal{G}$, for it contains two rotations about different axes. It will be shown that it can be readily reduced to an equivalent *RRRP* chain which is trivial, its degree of freedom being 1. In fact, this chain can first be decomposed into one containing only R and P pairs, namely, an *RRRRP* chain. Thus, $n = 3$ for the original chain, but $n' = 4$ for the decomposed one. The associated subgroups are $\mathcal{R}(A, \mathbf{e}_1)$, $\mathcal{R}(B, \mathbf{e}_1)$, $\mathcal{R}(C, \mathbf{e}_1)$, $\mathcal{R}(C, \mathbf{e}_2)$, and $\mathcal{P}(\mathbf{e}_2)$. Out of these subgroups, ten combinations of 2, ten of 3, five of 4 and one of 5 subgroups can be derived, i.e.,

$$N_2 = 10, \quad N_3 = 10, \quad N_4 = 5, \quad N_5 = 1$$

Out of the 2-subgroup products, only that of $\mathcal{R}(C, \mathbf{e}_2)$ and $\mathcal{P}(\mathbf{e}_2)$ is a subgroup of $\mathcal{G}$, namely $\mathcal{C}(C, \mathbf{e}_2)$. Moreover, out of the ten 3-subgroup products, those involving either the three R pairs of axis $\mathbf{e}_1$ or two R pairs of axis $\mathbf{e}_1$ and P produce the $\mathcal{E}(\mathbf{e}_1)$ subgroup, the remaining ones not being subgroups of $\mathcal{G}$. Furthermore, out of the 4-subgroup products, only that not involving $\mathcal{R}(C, \mathbf{e}_2)$ constitutes a subgroup, which is, again, $\mathcal{E}(\mathbf{e}_1)$. Clearly, the only one 5-subgroup product does not constitute a subgroup of $\mathcal{G}$. Thus, $\mathcal{G}_M = \mathcal{E}(\mathbf{e}_1)$, and the chain is exceptional. The intersection of the subgroups associated with the first three R pairs with $\mathcal{E}$ produces the same subgroup, $\mathcal{R}$, whereas the intersection of $\mathcal{C}(C, \mathbf{e}_2)$ with $\mathcal{E}(\mathbf{e}_1)$ produces $\mathcal{P}(\mathbf{e}_2)$, and hence, the reduced equivalent trivial chain is $RRRP$. For this, $d = 3$, $n = 3$, $r_0 = 2$, $r_1 = 2$, $r_2 = 2$, $r_3 = 2$. Hence, the degree of freedom, f, of the trivial chain, and hence, of the original exceptional chain, is 1, namely,

$$f = 3 \times 3 - 2 \times 4 = 1$$

Thus, the given $RRRC$ chain contains a redundant degree of freedom, namely, the rotational motion of the C pair. This redundant degree of freedom is present in virtually all linkages of the type under study, because of practical reasons that have to do with manufacturing and assembling.

As an additional example, consider the CCC chain proposed by Hervé (1978). This 3-link chain consists of three cylindrical pairs of coplanar axes, their associated subgroups being denoted by $\mathcal{C}(A, \mathbf{e}_1)$, $\mathcal{C}(B, \mathbf{e}_2)$, and $\mathcal{C}(C, \mathbf{e}_3)$, where $\mathbf{e}_1$, $\mathbf{e}_2$, and $\mathbf{e}_3$ are linearly dependent. Moreover, the product of the subgroups associated with the kinematic pairs of this chain does not constitute a subgroup of $\mathcal{G}$, as can be readily verified. It can be reduced to an equivalent PPP chain as follows: Decompose each pair into an R and a P pair, thereby obtaining an $RPRPRP$ chain of coplanar axes. Following the procedure outlined before, it is found that $\mathcal{G}_M = \mathcal{T}(\mathbf{e})$, where $\mathbf{e}$ is perpendicular to the plane of $\mathbf{e}_1$, $\mathbf{e}_2$, and $\mathbf{e}_3$. The intersection of the $\mathcal{C}$ subgroups of the original chain with $\mathcal{T}$ produces the subgroups $\mathcal{P}(\mathbf{e}_1)$, $\mathcal{P}(\mathbf{e}_2)$, and $\mathcal{P}(\mathbf{e}_3)$, whose product is $\mathcal{T}(\mathbf{e})$, and hence, the chain is exceptional. The equivalent PPP kinematic chain is trivial with $d = 2$, $n = 2$, $r_0 = r_1 = r_2 = 1$, and hence,

$$f = 2 \times 2 - 3 \times 1 = 1$$

Thus, the given CCC chain has a degree of freedom of 1.

Next an *RRRR* kinematic chain of skew axes is considered. Using the Chebyshev-Grübler-Kutzbach formula, this chain is shown to have a degree of freedom of -2, the negative sign indicating that it does not constitute a linkage, but rather a hyperstatic structure of redundance 2, which indicates that the structure is statically undetermined. However, Bennett (1903) showed that, for particular values of its link lengths and angles between its successive pair axes, the chain becomes a single-degree-of-freedom linkage, which is, since then, known as a *Bennett mechanism*. This is an example of a kinematic chain whose degree of freedom cannot be determined from its sole topology, i.e., it is elusive to the application of any Chebyshev-Grübler-Kutzbach formula.

As a matter of fact, one can show that any single, general closed kinematic chain, composed of R and P pairs whose degree of freedom is 1, has to have exactly 7 pairs, but, in all cases, a minimum of 4 rotational pairs. Indeed, under the foregoing assumptions, the chain is trivial, and the subgroups associated with all its pairs have a product which is the $\mathcal{G}$ group itself. Hence, in order to apply the Chebyshev-Grübler-Kutzbach formula to determine its degree of freedom, one has: $d = 6$, $n = 6, r_i = 5$, for $i = 0, 1, \ldots, 6$. Thus,

$$f = 6 \times 6 - 7 \times 5 = 1$$

Of course, any such chain having more than 3 P pairs, say $3 + p$, with $p > 0$, will have a value of m, as given by eq.(5.3.5), identical to p, and hence, its degree of freedom will be $1 - p$, which is less than 1. If the same chain is composed of only P pairs of noncoplanar axes, then the chain is trivial. the product of all the groups associated with its kinematic pairs being $\mathcal{U}(\mathbf{e}_1, \mathbf{e}_2, \mathbf{e}_3)$, and hence, $d = 3$ in the Chebyshev-Grübler-Kutzbach formula. Consequently, $r_i = 2$, for $i = 0, 1, \ldots, 6$. Moreover, in this case one can readily find that $m = 3$ and, hence, the following degree of freedom is derived:

$$f = 6 \times 3 - 7 \times 2 - 3 = 1$$

Several exceptions to the foregoing rule exist, which lead to paradoxical chains. One of these examples is the Bennett mechanism; another example is the *RRRRRR* chain with two triplets of concurrent axes. Such a chain has a single degree of freedom, which cannot be determined with the sole aid of its topology, and hence, another method should be resorted to. In the following section it is shown that the concept of *Jacobian* of a kinematic chain, containing information not only on the topology, but also on the geometry of the chain, allows the determination of its degree of freedom, even if the chain is of the paradoxical type. Moreover, it will be shown that this concept can be applied to determining the degree of freedom of complex kinematic chains.

The subject of paradoxical or overconstrained chains has attracted the attention of researchers over the years. It is worth mentioning in this regard the work of the French and the Australian schools, represented by Bricard (1927), Myard (1931), and Hervé in France, and Hunt (1978), Phillips (1984), and Baker—see, e.g., (Baker 1986) and the references therein—in Australia.

5.4 The Twist of a Link and the Jacobian of a Simple Kinematic Chain

The concept of twist, introduced in Chapter 3, is now recalled. A simple, open kinematic chain is considered, with only elementary kinematic pairs, i.e., either R or P. Clearly, if the chain originally contains lower kinematic pairs of other types, these can always be decomposed into a combination of only R and P pairs, as discussed in Section 5.2. Let O_i be a point on the axis of the elementary pair coupling the $(i-1)$st and the ith links, its velocity being denoted by $\mathbf{v}_i$. Moreover, the angular velocity of the ith link is denoted by ω_i. The twist of this link is defined as the 6-dimensional vector $\mathbf{t}_i$ given next:

$$\mathbf{t}_i \equiv \begin{bmatrix} \omega_i \\ \mathbf{v}_i \end{bmatrix} \tag{5.4.1}$$

Furthermore, if the ith pair is rotational, the relative motion of the $(i+1)$st link with respect to the ith one is defined by a rotation about the axis of the pair, whose direction is given in turn by the unit vector $\mathbf{e}_i$, through an angle θ_i. If the said pair is prismatic, then the same relative motion is defined by a displacement b_i along the direction of the unit vector $\mathbf{e}_i$. Furthermore, the vector directed from O_i to O_{i+1}, is denoted by $\mathbf{a}_i$, and a point P of the end link, the nth one, different from O_n, is defined arbitrarily. Vector $\mathbf{a}_n$ is defined as that directed from O_n to P. Finally, vectors $\mathbf{r}_i$, for $i = 1, \ldots, n$, are defined as those joining O_i with P, directed from the former to the latter, i.e.,

$$\mathbf{r}_i = \mathbf{a}_i + \mathbf{a}_{i+1} + \ldots + \mathbf{a}_n, \quad i = 1, \ldots, n \tag{5.4.2a}$$

If the chain contains only R pairs, the angular velocity, ω_n, of the nth link, is given by

$$\omega_n = \dot{\theta}_1 \mathbf{e}_1 + \ldots + \dot{\theta}_n \mathbf{e}_n \tag{5.4.2b}$$

The velocity of P is next obtained from the time derivative of its position vector. Let this vector be denoted by $\mathbf{p}$. Then, clearly,

$$\mathbf{p} = \mathbf{a}_1 + \mathbf{a}_2 + \ldots + \mathbf{a}_n \tag{5.4.3}$$

Hence,

$$\mathbf{v}_n \equiv \dot{\mathbf{p}} = \dot{\mathbf{a}}_1 + \dot{\mathbf{a}}_2 + \ldots + \dot{\mathbf{a}}_n \tag{5.4.4}$$

where,

$$\dot{\mathbf{a}}_k = \omega_k \times \mathbf{a}_k = (\dot{\theta}_1 \mathbf{e}_1 + \ldots + \dot{\theta}_k \mathbf{e}_k) \times \mathbf{a}_k, \quad k = 1, \ldots, n \tag{5.4.5}$$

Substitution of eq.(5.4.5) into eq.(5.4.4) leads to the following:

$$\begin{aligned}\dot{\mathbf{p}} =& \dot{\theta}_1 \mathbf{e}_1 \times (\mathbf{a}_1 + \mathbf{a}_2 + \ldots + \mathbf{a}_n) + \\ & \dot{\theta}_2 \mathbf{e}_2 \times (\mathbf{a}_2 + \ldots + \mathbf{a}_n) + \ldots + \dot{\theta}_n \mathbf{e}_n \times \mathbf{a}_n\end{aligned}$$

The quantities to the right of the $\times$ sign in the foregoing equation are readily identified as $\mathbf{r}_1, \ldots, \mathbf{r}_n$ of eq.(5.4.2a), and hence, $\dot{\mathbf{p}}$ takes on the form:

$$\dot{\mathbf{p}} = \dot{\theta}_1 \mathbf{e}_1 \times \mathbf{r}_1 + \ldots + \dot{\theta}_n \mathbf{e}_n \times \mathbf{r}_n \tag{5.4.6}$$

Thus, the twist of the nth link can be written as

$$\mathbf{t}_n = \mathbf{J}\dot{\theta} \tag{5.4.7a}$$

where $\mathbf{J}$ is the *Jacobian matrix* or, for the sake of brevity, the *Jacobian* of the kinematic chain. Its kth column, $\mathbf{j}_k$, is given next as a 6-dimensional vector, namely,

$$\mathbf{j}_k = \begin{bmatrix} \mathbf{e}_k \\ \mathbf{e}_k \times \mathbf{r}_k \end{bmatrix} \tag{5.4.7b}$$

If the kth pair is prismatic, then the foregoing column changes to

$$\mathbf{j}_k = \begin{bmatrix} \mathbf{0} \\ \mathbf{e}_k \end{bmatrix} \tag{5.4.7c}$$

Vector $\dot{\theta}$ is defined, in turn, as

$$\dot{\theta} = [\, \dot{\theta}_1, \dot{\theta}_2, \ldots, \dot{\theta}_n \,]^T \tag{5.4.8}$$

where, of course, if the jth pair is prismatic, then $\dot{\theta}_j$ is replaced by $\dot{b}_j$. Now, the degree of freedom of the chain, i.e., of the end link, is simply the dimension of the range of $\mathbf{J}$. Since $\mathbf{J}$ is configuration dependent, i.e., $\mathbf{J} = \mathbf{J}(\theta)$, the degree of freedom of the chain is also configuration dependent. Clearly, it cannot be greater than 6, i.e., if f denotes the said degree of freedom, one has

$$f \equiv \dim[\text{range}(\mathbf{J})] \leq 6 \tag{5.4.9}$$

Next, the determination of the degree of freedom of closed kinematic chains, of any of the types considered in Section 5.3, is undertaken. To this end, it is first assumed that the nth link is coupled to the 0th link via an

R pair, its associated joint variable being denoted by θ_{n+1}. Now, clearly, the twist of the nth link can be written, alternatively, as

$$\mathbf{t}_n = -\dot{\theta}_{n+1}\mathbf{j}_{n+1} = -\dot{\theta}_{n+1}\begin{bmatrix}\mathbf{e}_{n+1} \\ \mathbf{e}_{n+1} \times \mathbf{r}_{n+1}\end{bmatrix} \tag{5.4.10a}$$

where $\mathbf{e}_{n+1}$ is the unit vector along the axis of the R pair coupling the nth and the 0th links. Vector $\mathbf{r}_{n+1}$ is, then, the one joining point P with itself, and hence, it vanishes. Thus, the twist of the nth link can be finally written as

$$\mathbf{t}_n = -\dot{\theta}_{n+1}\begin{bmatrix}\mathbf{e}_{n+1} \\ \mathbf{0}\end{bmatrix} \tag{5.4.10b}$$

If the foregoing coupling takes place by means of a prismatic pair, then, clearly, $\mathbf{t}_n$, as given by eq.(5.4.10b), changes to

$$\mathbf{t}_n = -\dot{b}_{n+1}\begin{bmatrix}\mathbf{0} \\ \mathbf{e}_{n+1}\end{bmatrix} \tag{5.4.10c}$$

Now the Jacobian matrix of the closed chain is defined similarly to the one defined for the open chain, except that it is added an $(n+1)$st column $\mathbf{j}_{n+1}$, thereby rendering it a $6 \times (n+1)$ matrix. This allows the rewriting of eq.(5.4.7a) as

$$\mathbf{J}\dot{\theta} = \mathbf{0} \tag{5.4.11}$$

Clearly, the degree of freedom, f, of the closed kinematic chain, is simply the *nullity* of $\mathbf{J}$, i.e., the dimension of the nullspace of $\mathbf{J}$, $\mathcal{N}(\mathbf{J})$, and hence,

$$f = \dim[\mathcal{N}(\mathbf{J})] \tag{5.4.12}$$

It is pointed out that eq.(5.4.12) provides unambiguously the degree of freedom of the closed kinematic chain under study, regardless of its type, for, contrary to the usual Chebyshev-Grübler-Kutzbach formulae, it is based not only on the topology of the chain, but also on its geometry. Furthermore, the foregoing result was pointed out by Woo and Freudenstein (1970) using line geometry. As those researchers point out, a previous similar result was first reported by Bricard (1929), as applicable only to revolute pairs, whereas Voinea and Atanasiou (1962) derived the same result as applicable to screw pairs of identical pitches. Next, some examples taken from Angeles (1987-2) are presented that illustrate the applications of the foregoing result.

5.4.1 Example 1

Consider a spatial 4-link closed kinematic chain coupled by four R pairs. The conditions under which this chain has a single degree of freedom are next derived. This is done using the Jacobian matrix of the chain. The distances between the different R axes are denoted by a_i, for $i = 1, 2, 3, 4$. Moreover, the common perpendiculars between two consecutive axes intersect in consecutive pairs. Additionally, the angles between successive pair axes, α_i, for $i = 1, 2, 3, 4$, are arbitrary, and the unit vectors defining the directions of the pair axes are denoted by $\mathbf{e}_i$, for $i = 1, 2, 3, 4$.

Furthermore, let $\mathcal{P}_1$ be the common perpendicular to the fourth and the first pairs, $\mathcal{P}_i$, for $i = 2, 3, 4$, being defined likewise. Hence, $\mathcal{P}_1$, $\mathcal{P}_2$ and the axis of the first R pair intersect at a common point O_1, points O_i, for $i = 2, 3, 4$, being similarly defined. Now, vectors $\mathbf{r}_1, \mathbf{r}_2, \mathbf{r}_3$ are defined as those directed from O_1, O_2 and O_3 to O_4, respectively. Then, the Jacobian matrix of the chain takes on the following form:

$$\mathbf{J} = \begin{bmatrix} \mathbf{e}_1 & \mathbf{e}_2 & \mathbf{e}_3 & \mathbf{e}_4 \\ \mathbf{e}_1 \times \mathbf{r}_1 & \mathbf{e}_2 \times \mathbf{r}_2 & \mathbf{e}_3 \times \mathbf{r}_3 & \mathbf{0} \end{bmatrix} \tag{5.4.13}$$

Therefore, the conditions under which the degree of freedom of this chain is 1 are those under which $\mathbf{J}$ becomes of rank 3. If $\mathbf{J}$ is thought of as a 6×4 matrix, which is given in two 3×4 blocks, clearly, the three rows of its upper block are linearly dependent. Moreover, since the three lower rows of the fourth column of $\mathbf{J}$ are zero, the conditions under which rank($\mathbf{J}$)= 3 are those under which the first three columns of the lower block are linearly dependent. Thus, the nullity of $\mathbf{J}$ is unity if the three vectors $\mathbf{e}_1 \times \mathbf{r}_1, \mathbf{e}_2 \times \mathbf{r}_2$ and $\mathbf{e}_3 \times \mathbf{r}_3$ are coplanar, a condition that can be expressed as

$$(\mathbf{e}_1 \times \mathbf{r}_1) \times (\mathbf{e}_2 \times \mathbf{r}_2) \cdot (\mathbf{e}_3 \times \mathbf{r}_3) = 0 \tag{5.4.14}$$

Next, let $\mathbf{f}_1$ be the unit vector directed from O_4 to O_1, vectors $\mathbf{f}_2$, $\mathbf{f}_3$, and $\mathbf{f}_4$ being similarly defined. From the geometry of the chain, one has

$$\mathbf{f}_{i+1} \times \mathbf{f}_i = A_{i+1}\mathbf{e}_{i+1};\ i = 1, \ldots, 4;\ i+1 = 1, \text{ if } i = 4 \tag{5.4.15a}$$

Of course,

$$A_{i+1} = \sin(\mathbf{f}_{i+1}, \mathbf{f}_i) \tag{5.4.15b}$$

On the other hand,

$$\mathbf{e}_{i+1} \times \mathbf{e}_i = \mathbf{f}_i \sin \alpha_i;\ i = 1, \ldots, 4;\ i+1 = 1, \text{ if } i = 4 \tag{5.4.16}$$

The closure of the chain leads to

$$\sum_1^4 a_i \mathbf{f}_i = \mathbf{0} \tag{5.4.17}$$

Moreover, vectors $\mathbf{r}_i$, for $i = 1, 2, 3, 4$, can be written as

$$\mathbf{r}_1 = -a_4\mathbf{f}_4, \quad \mathbf{r}_2 = -a_4\mathbf{f}_4 - a_1\mathbf{f}_1, \quad \mathbf{r}_3 = a_3\mathbf{f}_3 \tag{5.4.18}$$

Upon substitution of eqs.(5.4.18) into eq.(5.4.14), the following is obtained:

$$\mathbf{p}\cdot\mathbf{f}_3 = 0 \tag{5.4.19a}$$

where vector $\mathbf{p}$ is defined as

$$\mathbf{p} = [-a_1a_4(\mathbf{e}_1\times\mathbf{f}_4\cdot\mathbf{f}_1)\mathbf{e}_2 + \mathbf{e}_1\times a_4\mathbf{f}_4\cdot\mathbf{e}_2(a_1\mathbf{f}_1 + a_4\mathbf{f}_4)]\times\mathbf{e}_3 \tag{5.4.19b}$$

Next, substitution of eqs.(5.4.15 & 18) into eq.(5,4,19b) leads to

$$a_1A_1s\alpha_2(\mathbf{f}_2\cdot\mathbf{f}_3) + a_2A_3s\alpha_1(\mathbf{f}_4\cdot\mathbf{f}_1) = 0 \tag{5.4.20}$$

where $c(\cdot)$ and $s(\cdot)$ denote $\cos(\cdot)$ and $\sin(\cdot)$, respectively. On the other hand, the inner product of both sides of eq.(5.4.17) by vector $\mathbf{f}_1\times\mathbf{f}_2$ yields

$$a_3\mathbf{f}_3\cdot\mathbf{f}_1\times\mathbf{f}_2 + a_4\mathbf{f}_4\cdot\mathbf{f}_1\times\mathbf{f}_2 = 0 \tag{5.4.21}$$

Next, vectors $\mathbf{f}_1s\alpha_1$ and $\mathbf{f}_2s\alpha_2$, as given by eqs.(5.4.15), are substituted into eqs.(5.4.16), which yields

$$\mathbf{f}_4\cdot\mathbf{f}_1\times\mathbf{f}_2 = -A_1A_2s\alpha_1, \quad \mathbf{f}_3\cdot\mathbf{f}_1\times\mathbf{f}_2 = -A_2A_3s\alpha_2 \tag{5.4.22}$$

The substitution of eqs.(5.4.22) into eq.(5.4.21) yields

$$A_1a_4s\alpha_1 + A_3a_3s\alpha_2 = 0 \tag{5.4.23}$$

Moreover, substitution of eq.(5.4.23) into eq.(5.4.20) leads to

$$a_1a_3s^2\alpha_2(\mathbf{f}_2\cdot\mathbf{f}_3) = a_2a_4s^2\alpha_1(\mathbf{f}_4\cdot\mathbf{f}_1) \tag{5.4.24}$$

The inner products of eq.(5.4.24) are now written in terms of vectors $\mathbf{e}_i$, for $i = 1,2,3,4$, using eqs.(5.4.16), thereby obtaining

$$\mathbf{f}_2\cdot\mathbf{f}_3 = \frac{(\mathbf{e}_2\times\mathbf{e}_3)\times\mathbf{e}_3\cdot\mathbf{e}_4}{s\alpha_2s\alpha_3}, \quad \mathbf{f}_4\cdot\mathbf{f}_1 = \frac{(\mathbf{e}_2\times\mathbf{e}_1)\times\mathbf{e}_1\cdot\mathbf{e}_4}{s\alpha_1s\alpha_4} \tag{5.4.25}$$

Finally, substitution of eqs.(5.4.25) into eq.(5.4.24) leads to

$$a_1a_3s\alpha_2s\alpha_4(\mathbf{e}_2\times\mathbf{e}_3)\times\mathbf{e}_3\cdot\mathbf{e}_4 = a_2a_4s\alpha_1s\alpha_3(\mathbf{e}_2\times\mathbf{e}_1)\times\mathbf{e}_1\cdot\mathbf{e}_4 \tag{5.4.26a}$$

Upon expansion of the vector products involved,

$$a_1a_3s\alpha_2s\alpha_4(c\alpha_2c\alpha_3 - \mathbf{e}_2\cdot\mathbf{e}_4) = a_2a_4s\alpha_1s\alpha_3(c\alpha_1c\alpha_4 - \mathbf{e}_2\cdot\mathbf{e}_4) \tag{5.4.26b}$$

which is the desired condition. This equation holds if the two following equations hold as well:

$$a_1a_3s\alpha_2s\alpha_4 = a_2a_4s\alpha_1s\alpha_3, \quad c\alpha_2c\alpha_3 = c\alpha_1c\alpha_4 \tag{5.4.27}$$

It is pointed out that eqs.(5.4.27) were derived using a different approach by Ho (1978). Next, the conditions under which eq.(5.4.27) holds are discussed:

i) $\alpha_1 = \alpha_2 = \alpha_3 = \alpha_4 = 0$. This condition leads to planar four-link mechanisms.

ii) $a_1 = a_2 = a_3 = a_4 = 0$. This condition leads to spherical four-link mechanisms.

iii) $a_1 = a_3, a_2 = a_4, s\alpha_1 = s\alpha_3, s\alpha_2 = s\alpha_4, c\alpha_1 = c\alpha_3, c\alpha_2 = c\alpha_4$, i.e., $\alpha_1 = \alpha_3$, $\alpha_2 = \alpha_4$ and $a_1s\alpha_2 = a_2s\alpha_1$. This condition leads to the Bennett mechanisms.

5.4.2 Example 2

The degree of freedom of the $PRPRPR$ equivalent of the CCC closed chain with kinematic pairs of coplanar axes, already discussed, is derived using its Jacobian matrix. Let **f**, **g**, and **h** denote the unit vectors defining the directions of the C pairs of the original chain, which are, in turn, the directions of the axes of each set of consecutive P and R pairs of the equivalent 6-link chain under study. Clearly, the three foregoing unit vectors are coplanar. Moreover, let O_1, O_2, and O_3 be the intersections of each pair of revolute axes of the chain, and let **u** and **v** denote the vectors directed from O_1 and O_2 to O_3, respectively. The Jacobian matrix of the chain, thus, is the following:

$$\mathbf{J} = \begin{bmatrix} \mathbf{f} & \mathbf{0} & \mathbf{g} & \mathbf{0} & \mathbf{h} & \mathbf{0} \\ \mathbf{0} & \mathbf{f} & \mathbf{g} \times \mathbf{u} & \mathbf{g} & \mathbf{0} & \mathbf{h} \end{bmatrix}$$

Application of some elementary transformations to **J**, not altering its rank, lead to the following form:

$$\mathbf{J}' = [\mathbf{J}_1 \quad \mathbf{J}_2]$$

with submatrices $\mathbf{J}_1$ and $\mathbf{J}_2$ defined as

$$\mathbf{J}_1 = \begin{bmatrix} \mathbf{0} & \mathbf{0} & \mathbf{0} \\ \mathbf{f} & \mathbf{g} & \mathbf{h} \end{bmatrix}, \quad \mathbf{J}_2 = \begin{bmatrix} \mathbf{f} & \mathbf{g} & \mathbf{h} \\ \mathbf{0} & \mathbf{g} \times \mathbf{u} & \mathbf{0} \end{bmatrix}$$

Now, since each column of $\mathbf{J}_1$ is linearly independent with the columns of $\mathbf{J}_2$, and vice versa, one has

$$\text{rank}(\mathbf{J}) = \text{rank}(\mathbf{J}_1) + \text{rank}(\mathbf{J}_2)$$

Moreover, the rank of $\mathbf{J}_1$ is 2 and that of $\mathbf{J}_2$ is 3, the rank of **J** thus being 5; therefore, the nullity of **J** is 1, and hence, the degree of freedom of the chain is 1, a result previously obtained using group theory.

5.5 The Condition Number of a Kinematic Chain. Isotropy

In this section, the condition number, κ, of a simple, open kinematic chain is defined as that of its Jacobian matrix. This is defined, in turn, for an $n \times n$ matrix **J** as (Golub and Van Loan 1983)

$$\kappa(\mathbf{J}) \equiv \|\mathbf{J}\| \, \|\mathbf{J}^{-1}\| \tag{5.5.1}$$

where $\|\cdot\|$ denotes any norm of its matrix argument. The condition number of nonsquare matrices can also be defined, but in this chapter only square Jacobian matrices will be considered, and hence the foregoing definition will suffice.

Following the invariant approach adopted from the outset, it is only natural to define a frame-invariant condition number. This is readily obtained if the Euclidean norm is adopted in eq.(5.5.1), namely

$$\| \mathbf{J} \|_W \equiv \sqrt{\mathrm{tr}(\mathbf{J}^T \mathbf{W} \mathbf{J})} \tag{5.5.2}$$

where $\mathbf{W}$ is an $n \times n$ positive definite weighting matrix that is introduced for purposes of normalization. For instance, if $\mathbf{W}$ is defined as $(1/n)\mathbf{1}$, then the Euclidean norm of the $n \times n$ identity matrix turns out to be unity. In the discussion that follows, only the Euclidean norm will be considered, and hence, the subscript W will be dropped from the norm symbol. The significance of the condition number of $\mathbf{J}$ will be discussed.

Clearly, from the definition introduced in eq.(5.4.7a), $\mathbf{J}$ is square if $n = 6$. In this case, given a prescribed twist $\mathbf{t}_n$ of the end link, eq.(5.4.7a) can be solved for the vector of joint rates, $\dot{\boldsymbol{\theta}}$, uniquely if $\mathbf{J}$ is nonsingular. When solving for the said vector with finite precision, roundoff errors are invariably introduced into the data, $\mathbf{J}$ and $\mathbf{t}_n$. Hence, the result will invariably be affected by roundoff error as well. The condition number of $\mathbf{J}$ is the amplification factor of relative roundoff errors in the results, arising from relative roundoff errors in the data. Furthermore, from the definition of the condition number, it ranges from 1 to ∞, i.e.,

$$1 \leq \kappa < \infty \tag{5.5.3}$$

Thus, the minimum value that the condition number of a matrix can attain is 1. Matrices with a condition number of 1 have all their proper values on a circle of radius a centered at the origin of the complex plane, and hence, are called *isotropic*. Notice that isotropic matrices are multiples of orthogonal matrices.

Since the Jacobian matrix of a kinematic chain is configuration dependent, its condition number is similarly configuration dependent. Thus, for a given kinematic chain, the condition number varies between a minimum value, say κ_m, and ∞. If $\kappa_m = 1$, the manipulator is said to be *isotropic*. The Jacobian matrix of a square matrix attains infinite values when the matrix becomes singular, which is obvious from the definition introduced in eq.(5.5.1). Before one continues the discussion on isotropic manipulators, it is helpful to introduce the following:

Theorem 5.5.1: (Angeles and Rojas 1987) *The frame-invariant condition number of a general, simple, open kinematic 7-link chain is independent from the variables associated with the first and the last pairs.*

The proof of this Theorem follows from the fact that the first and the last pairs, whether R or P, produce a rigid-body motion of the overall chain, which hence does not affect its frame-invariant condition number. Moreover, the theorem applies to general kinematic chains, whose end link is capable of attaining arbitrary positions and orientations. A planar 3-link chain is not capable of orienting its end link arbitrarily in the 3-dimensional space, its set of orientations being constrained to take place about an axis that is normal to the chain's plane. Hence, the only joint variable not affecting its frame-invariant condition number is that associated with its first pair. However, Theorem 5.5.1 applies as such to 3-link spherical kinematic chains. Indeed, the motions undergone by the end link of these chains form the spherical group, and hence, a rotation of the first or the third joints is equivalent to a rigid-body rotation of the overall chain, which naturally does not affect its frame-invariant condition number.

As a consequence of the foregoing, the set of configurations of an isotropic kinematic chain, at which its condition number attains its minimum value of 1, are not isolated, but form an infinite, non denumerable set. Consider, for instance, an isotropic 3-link planar kinematic chain, whose two R pairs are centered at points O_1 and O_2, the end point of the end link being denoted by P. Moreover, let the distances from O_1 to O_2 and from O_2 to P be denoted by a_1 and a_2, respectively. As shown in Salisbury and Craig (1982), the ratio $\alpha = a_2/a_1$ producing such an isotropic chain is $\sqrt{2}/2$. It is interesting to notice that many 3-link kinematic chains appearing in the human body, such as the radius-humerus, the femur-tibia sets, and the sets of consecutive falanxes, observe a ratio approximately equal to the foregoing value. In fact, Leonardo da Vinci, a pioneer anthropometrist, points out that the link-length ratio between the humerus and the radius is, subject to natural variations from specimen to specimen, a quantity around 71.4% (Schuman 1952). For such a chain, the set of configurations of minimum condition number, $\kappa = 1$, is defined by

$$\theta_1 = \text{arbitrary}, \quad \theta_2 = \pm 3\pi/4$$

where θ_2 is the angle measured from the line defined by segment O_1O_2 to segment O_2P. Hence, all points lying on a circle centered at O_1, of radius $\sqrt{2}a_1/2$, are isotropic. This circle is, thus, called *the isotropy circle* of the given chain.

As a second example, a 4-link spherical kinematic chain is considered. This chain is composed of three R pairs of axes intersecting at a common point O. Let the angle between the axes of the first two pairs be denoted by α_1, that between the last two axes by α_2. As shown in Angeles and Rojas (1987), the chain under study is isotropic if $\alpha_i = \pi/2$, for $i = 1,2$. Moreover, the associated condition number attains its minimum value of 1 when the planes defined by the first two and the last two pair axes form a right dihedron. That is, if the said angle is denoted by θ_2, isotropy occurs when $\theta_2 = \pm\pi/2$. In this case, the set of isotropic points can be determined by resorting to the concept of *workspace* of the kinematic chain. This is defined as the set of all configurations of the chain that are kinematically possible. In the case of the planar 3-link chain, of arbitrary link lengths a_1 and a_2, the said workspace is an annulus of greater radius $a_1 + a_2$ and smaller radius $|a_1 - a_2|$. Alternatively, the workspace can be described in terms of the joint variables θ_1 and θ_2, namely as $-\pi \leq \theta_1, \theta_2 \leq \pi$, which is a square centered at the origin of the θ_1, θ_2 plane, of sides parallel to the θ_1 and θ_2 axes, and lengths 2π. Thus, the chain under study maps the square into the annulus. Similarly, the workspace of the 4-link spherical chain can be readily described in terms of its joint variables, as $-\pi \leq \theta_i \leq \pi$, for $i = 1, 2, 3$. The same workspace can be described in terms of the coordinates defining the orientation of the end link as follows: Let the 4-dimensional spherical surface $\mathcal{K}: \lambda^T\lambda = 1$ represent the set of all possible configurations that the unconstrained end link can attain when rotating about point O, where λ is the 4-dimensional vector of linear invariants of the end link. Because of kinematic couplings, only a subset $\mathcal{K}' \subseteq \mathcal{K}$ of configurations can be attained. Thus, the workspace of the chain is a part of the 4-dimensional unit spherical surface. Moreover, if $\mathbf{e}_1$, $\mathbf{e}_2$, and $\mathbf{e}_3$ denote the unit vectors parallel to the axes of the R pairs of the chain, then the locus of isotropic configurations is defined by $\theta_2 = \pm\pi/2$, and hence, the four coordinates of a point in the space of configurations that can be obtained by the unconstrained end link are subjected to two constraints when the chain is in an isotropic configuration, namely, the one defining the unit magnitude of its position vector and the one defining its isotropy. Thus, the locus of isotropic configurations of such a chain is a two-dimensional surface imbedded in the aforementioned 4-dimensional space. Then, whereas the locus of isotropic points of a 3-link planar chain is a curve, in fact, a circle, that of the 4-link spherical chain is a surface whose geometry is more complex than that of the former. The detailed derivation of the algebraic description of this surface is reported in (Angeles 1988). In the same reference, the surface is displayed in both the space of linear invariants and that of Euler parameters. For completeness, the locus of singular points of the same chain is derived in the above reference. It is shown there that the said locus is a curve, rather than a surface.

Finally, an isotropic $6R$, spatial, open kinematic chain has been proposed by Angeles and López-Cajún (1988). In this chain, all rotational axes make right angles and lie a distance a apart. Moreover, the common perpendiculars of every two successive pairs are offset a distance a. In this case, the set of configurations of the end link is a portion of the 6-dimensional space of link screws. The set of isotropic points of this chain is a 4-dimensional surface whose projection onto the 3-dimensional Euclidean space is a torus, a result of the fact that the rotations associated with the first and the last pairs do not affect the condition number of the chain, and hence, its isotropy.

5.6 Complex Kinematic Chains. An Overview

A kinematic chain containing links coupled to three or more links is termed *complex*. In this regard, a link coupled to only one other link is referred to as *single*; if coupled to two other links, as binary. Ternary and quaternary links are similarly defined. In general, a link coupled to k other links will be said to be *of connectivity* k. Clearly, any complex kinematic chain can be regarded as the coupling of simple kinematic chains, and hence, one can always extract simple kinematic chains from a complex one. Thus, any subset of coupled links of a complex chain is a chain itself, and hence, will be termed a *subchain* of the given complex chain. A subchain can in turn be either simple or complex.

A complex kinematic chain may or may not contain closed subchains. If it does not, then it is said to have a *tree structure*. However, even if a complex chain is not of the tree-structure type, it can always be subdivided into sets of tree-structure chains. Some methods of dynamical analysis of mechanical systems are based on tree-structure kinematic chains (Wittenburg 1977). However, in this study, this type of chain is not of any particular interest, and hence, will be omitted from this discusion.

If a complex kinematic chain contains open subchains of the simple type, then the overall chain can be regarded as composed of two subsets, one being a kinematic chain composed only of closed subchains, the other being a set of open, simple subchains. It will become apparent that the analysis of the complex chain is reduced to that of the subchain containing only closed subchains. Moreover, each of these closed subchains is termed a *loop*. Given the relevance of complex kinematic chains containing only closed subchains, these will be termed *multiple-loop kinematic chains*. Introduced in Section 5.7 is a method to determine the degree of freedom of this type of chain. A detailed discussion of complex kinematic chains is given in (Gosselin 1988).

5.7 The Jacobian of a Multiple-Closed-Loop Kinematic Chain. Degree of Freedom

In some instances, the Chebyshev-Grübler-Kutzbach formula introduced in Section 5.3 suffices to determine the degree of freedom of a multiple-closed-loop kinematic chain. For instance, a *Stewart platform* (Stewart 1965), usually found as a flight simulator, is a kinematic chain composed of 32 links coupled as follows: Two of these are of connectivity 6, i.e., they are coupled to six links, the remaining links being binary. One of the links of connectivity 6 is arbitrarily termed *the base*, the other being the *moving platform*. Moreover, the base and the moving platform are coupled to each other via six *legs*, each of which is a five-link open chain of the simple type—the aforementioned five links become intermediate links located between the base and the platform. Thus, the Stewart platform can be regarded as being composed of six open seven-link kinematic subchains of the simple type. In general, each kinematic pair of this chain can be either R or P. Thus, the chain contains a total of 36 kinematic pairs of either one of the two aforementioned types. Since each of these pairs imposes 5 constraints, application of the Chebyshev-Grübler-Kutzbach formula to this chain leads to

$$l = 31 \times 6 - 36 \times 5 = 6$$

and hence, the chain can reproduce the arbitrary motions undergone by a rigid body—e.g., an airplane—in the 3-dimensional space. In the foregoing, the degree of freedom of the chain is understood, of course, as that of the moving platform.

A far simpler kinematic chain of the complex type is the *double parallelogram*, which is frequently cited as an example elusive to Chebyshev-Grübler-Kutzbach formulae. This is a planar chain composed of two ternary and three binary links, coupled by revolute pairs of parallel axes. Moreover, the binary links are all of the same length and the distances between axes coupling each ternary link to the binary ones is the same. Thus, the ternary links move with pure relative translation with respect to each other, and the degree of freedom of the chain is 1. However, application of the Chebyshev-Grübler-Kutzbach formula to this chain produces an incorrect degree of freedom. Indeed, since the chain is composed of 5 links and 6 revolute pairs, application of the aforementioned formula leads to the following:

$$l = (5 - 1) \times 3 - 6 \times 2 = 0$$

The Chebyshev-Grübler-Kutzbach formula is not applicable in this case because it does not consider the geometry of the chain, only its topology. What the formula states is that a chain with the given topology and *arbitrary* dimensions is an isostatic planar structure. The determination of the degree of freedom of a multiple-closed-loop kinematic chain, based on its *Jacobian*, is next undertaken. To this end, the chain is first divided into i *independent* closed subchains of the simple type. These are readily determined resorting to *graph theory* (Harary 1972). In this context, a graph is assigned to the chain, in which a *node* represents a link, and an *edge* represents a kinematic pair of either type, R or P. If the chain is composed of n links and p such pairs, then, application of Euler's formula for a connected graph (Harary 1972) to the resulting graph leads to the following number of *independent loops*, i.e., of *independent closed subchains* of the simple type:

$$i = p - n + 1 \tag{5.7.1}$$

Next, a vector θ is defined containing the variables associated with each of the kinematic pairs, its time derivative being $\dot{\theta}$. Clearly, the two foregoing vectors are p-dimensional. Now, the following kinematic closure equation is written for each closed subchain following the procedure introduced in Section 5.4 for simple chains:

$$\mathbf{J}_k\dot{\theta} = \mathbf{0}, \quad k = 1, \ldots, i \tag{5.7.2}$$

where $\mathbf{J}_k$ is the Jacobian matrix of the kth closed subchain, also introduced in Section 5.4. The i foregoing equations can now be assembled and written as a single vector equation of a higher dimension. To this end, the Jacobian matrix of the overall multiple-closed-loop kinematic chain, $\mathbf{J}$, is defined as:

$$\mathbf{J} \equiv \begin{bmatrix} \mathbf{J}_1 \\ \mathbf{J}_2 \\ \vdots \\ \mathbf{J}_i \end{bmatrix} \tag{5.7.3}$$

Thus, the i equations comprised in eq.(5.7.2) can now be written as

$$\mathbf{J}\dot{\theta} = \mathbf{0} \tag{5.7.4}$$

Clearly, the degree of freedom of the chain under study, l, can be readily determined as the nullity of $\mathbf{J}$, i.e.,

$$l = \dim(\mathcal{N}) \tag{5.7.5}$$

where $\mathcal{N}$ represents the nullspace of $\mathbf{J}$. Determining the said nullity is a simple problem of linear algebra that, due to the possible high dimension of matrix $\mathbf{J}$, may require one to resort to methods of numerical analysis. In any instance, it is a straightforward problem. As an example of the application of the foregoing concepts, the degree of freedom of the double prallelogram is next determined:

As described above, the double parallelogram is a five-link kinematic chain with two closed loops. It contains three binary and two ternary links. Moreover, the axes of the rotational pairs of each ternary link are equally spaced, the spacing being a in the two links. The lengths of the binary links, on the other hand, are identical and equal to b, and hence, the two ternary links move with relative pure translation (Angeles 1987-2). Let one of the two ternary links be termed the *fixed link*, the other being the *coupler link*. Furthermore, let the three angles defining the rotations associated with each rotational pair of the fixed link be successively denoted by θ_1, θ_2, and

θ_3, those of the coupler link by θ_4, θ_5, and θ_6. The time derivatives of these variables are denoted by $\dot{\theta}_i$, for $i = 1, \ldots, 6$. Let $\mathbf{J}_1$ denote the Jacobian matrix of the first loop, $\mathbf{J}_2$ denoting the Jacobian matrix of the second one. This allows one to arrive at the following relations:

$$\mathbf{J}_1\dot{\theta} = \mathbf{0}, \quad \mathbf{J}_2\dot{\theta} = \mathbf{0} \tag{5.7.6}$$

where $\mathbf{J}_1$ and $\mathbf{J}_2$ are defined as follows:

$$\mathbf{J}_1 = \begin{bmatrix} \mathbf{e} & \mathbf{e} & \mathbf{e} & \mathbf{e} & \mathbf{0} & \mathbf{0} \\ \mathbf{e}\times\mathbf{r}_1 & \mathbf{e}\times\mathbf{r}_2 & \mathbf{e}\times\mathbf{r}_3 & \mathbf{0} & \mathbf{0} & \mathbf{0} \end{bmatrix} \tag{5.7.7a}$$

$$\mathbf{J}_2 = \begin{bmatrix} \mathbf{0} & \mathbf{0} & \mathbf{e} & \mathbf{e} & \mathbf{e} & \mathbf{e} \\ \mathbf{0} & \mathbf{0} & \mathbf{e}\times\mathbf{r}_3 & \mathbf{0} & \mathbf{e}\times\mathbf{s}_1 & \mathbf{e}\times\mathbf{s}_2 \end{bmatrix} \tag{5.7.7b}$$

the unit vector $\mathbf{e}$ being perpendicular to the plane of the chain. Vectors $\mathbf{r}_1$, $\mathbf{r}_2$, and $\mathbf{r}_3$ are defined as those directed to one point of the axis of one of the fixed pairs of the first loop, from a point of the axis of each of the other three pairs of the same loop. Vectors $\mathbf{s}_1$, $\mathbf{s}_2$, and $\mathbf{s}_3$ are defined likewise for the second loop. Thus, the Jacobian matrix of the entire chain is the following:

$$\mathbf{J} = \begin{bmatrix} \mathbf{J}_1 \\ \mathbf{J}_2 \end{bmatrix}$$

The dimensions of the chain permit the writing of the following:

$$\mathbf{r}_3 = \mathbf{r}_2 - \mathbf{r}_1, \quad \mathbf{s}_1 = -\mathbf{r}_1, \quad \mathbf{s}_2 = \mathbf{r}_2 - 2\mathbf{r}_1$$

Substitution of the foregoing relations into the expression given for $\mathbf{J}$, and further application of some elementary transformations not affecting the rank of $\mathbf{J}$, leads to the following transformed Jacobian, $\mathbf{J}'$:

$$\mathbf{J}' = \begin{bmatrix} \mathbf{e} & \mathbf{e} & \mathbf{e} & \mathbf{e} & \mathbf{0} & \mathbf{0} \\ \mathbf{e}\times\mathbf{r}_1 & \mathbf{e}\times\mathbf{r}_2 & \mathbf{0} & \mathbf{0} & \mathbf{0} & \mathbf{0} \\ \mathbf{0} & \mathbf{0} & \mathbf{0} & \mathbf{e} & \mathbf{e} & \mathbf{0} \\ \mathbf{0} & \mathbf{0} & \mathbf{e}\times\mathbf{r}_2 & \mathbf{0} & -\mathbf{e}\times\mathbf{r}_1 & \mathbf{0} \end{bmatrix}$$

from which it is apparent that the rank of $\mathbf{J}'$, and hence, that of $\mathbf{J}$, is 5, its nullity thus being 1, which is the degree of freedom of the chain.

5.8 Holonomic and Nonholonomic Constraints

In this section an outline is given of the treatment of kinematic constraints, whether produced by a lower kinematic pair or any other holonomic constraint, or even by nonholonomic constraints. First, the extension to holonomic constraints other than the lower kinematic pairs is introduced. Holonomic constraints owe their name to the greek word holos, which means *integer*, and denotes constraints that can be described in *integral form*, i.e., in terms of displacements, as opposed to *differential form*, i.e., in terms of velocities—including angular velocities. From the results of Chapter 3 it is apparent that the angular-velocity vector is not a total derivative, i.e., no vector exists whose time derivative is the angular-velocity vector. The foregoing is true, unless the angular velocity takes place about a fixed axis, which happens when the motion is constrained to take place in a plane, thereby reducing itself to *planar motion*. As a consequence, unless the motion is planar, any kinematic constraint of the differential type that involves the angular velocity of a rigid body is nonholonomic. For instance, consider a rigid sphere rolling without slipping on a certain plane. Let $\mathbf{c}$ and $\mathbf{q}$ denote the position vectors of its centroid C and of the contact point Q between the sphere and the plane. The non-slip conditions can thus be established by relating the velocity of C, $\dot{\mathbf{c}}$ with the angular velocity ω of the sphere, which can be stated as

$$\dot{\mathbf{c}} = \omega \times (\mathbf{c} - \mathbf{q})$$

The foregoing equation cannot be written in integral form for, although its left-hand side can be integrated to produce $\mathbf{c} + \mathbf{k}$, where $\mathbf{k}$ is an integration constant vector, the right-hand side cannot be integrated to yield an expression in the rotation tensor and the position vectors of C and Q, free of time derivatives.

Mechanical couplings that produce holonomic constraints other than the lower kinematic pairs are, among others, pulley-belt, sprocket-chain, and cam-follower transmissions, as well as gear trains. These couplings, and those associated with the lower kinematic pairs, produce constraints on the twists of the coupled bodies, say $\mathbf{t}_{i-1}$ and $\mathbf{t}_i$, of the following form:

$$\mathbf{A}_{i,i-1}\mathbf{t}_{i-1} + \mathbf{A}_{i,i}\mathbf{t}_i = \mathbf{0} \tag{5.8.1a}$$

where the twists are now defined as follows:

$$\mathbf{t}_i \equiv \begin{bmatrix} \omega_i \\ \dot{\mathbf{c}}_i \end{bmatrix} \tag{5.8.1b}$$

and $\dot{\mathbf{c}}_i$ is the velocity of the center of mass of the ith rigid body of the system. The foregoing concepts are next illustrated with a few examples.

For instance, if the two bodies are coupled by a revolute pair, then the 6×6 matrices $\mathbf{A}_{i,i-1}$ and $\mathbf{A}_{i,i}$, of eq.(5.8.1), are the following:

$$\mathbf{A}_{i,i-1} = \begin{bmatrix} \mathbf{1} \times \mathbf{e}_i & \mathbf{0} \\ \mathbf{1} \times \rho_i & \mathbf{1} \end{bmatrix}, \quad \mathbf{A}_{i,i} = \begin{bmatrix} \mathbf{1} \times \mathbf{e}_i & \mathbf{0} \\ \mathbf{1} \times (\mathbf{a}_i + \rho_i) & \mathbf{1} \end{bmatrix} \tag{5.8.2a}$$

where $\mathbf{a}_i$, $\mathbf{e}_i$, and ρ_i denote the vector directed from a point O_{i-1} of the axis of the $(i-1)$st R pair to a point O_i of the axis of the ith R pair, a unit vector parallel to the axis of the ith R pair, and the vector directed from O_i to the center of mass of the ith link. Moreover, $\mathbf{0}$ and $\mathbf{1}$ denote the zero and the identity Cartesian tensors.

Now, if the coupling takes place via a prismatic pair, then, the aforementioned matrices take on the following forms:

$$\mathbf{A}_{i,i-1} = \begin{bmatrix} \mathbf{1} & \mathbf{0} \\ \mathbf{0} & \mathbf{1} \times \mathbf{e}_i \end{bmatrix}, \quad \mathbf{A}_{i,i} = \begin{bmatrix} \mathbf{1} & \mathbf{0} \\ (\mathbf{1} \times \mathbf{e}_i)(\mathbf{1} \times \mathbf{a}_i) & \mathbf{1} \times \mathbf{e}_i \end{bmatrix} \tag{5.8.2b}$$

where all variables are defined in exactly the same way as for the previous case, except that $\mathbf{e}_i$ is now defined properly as the unit vector parallel to the direction of relative translation along the ith joint, that is now prismatic.

Furthermore, if the coupling takes place via a gear train composed of, say, two bevel gears of axes making an angle of α and parallel to the unit vectors $\mathbf{e}_{i-1}$ and $\mathbf{e}_i$, then,

$$\mathbf{A}_{i,i-1} = \begin{bmatrix} \mathbf{F} & \mathbf{0} \\ \mathbf{H} & \mathbf{G} \end{bmatrix}, \quad \mathbf{A}_{i,i} = \begin{bmatrix} -\mathbf{F} & \mathbf{0} \\ \mathbf{0} & -\mathbf{G} \end{bmatrix} \tag{5.8.2c}$$

where tensor $\mathbf{F}$ is defined as

$$\mathbf{F} \equiv \frac{\partial(\mathbf{f} \times \mathbf{v})}{\partial \mathbf{v}} \tag{5.8.3a}$$

and the unit vector $\mathbf{f}$ is directed along the common element of the two conic pitch surfaces of the bevel gears under study. Moreover, if ρ_1 and ρ_2 are the vectors locating the mass centers of the two gears from the point of intersection of their two axes, i.e., from the common apex of the two pitch conical surfaces, then tensors $\mathbf{G}$ and $\mathbf{H}$ are the following:

$$\mathbf{G} \equiv (\rho_2 - \rho_2) \otimes \mathbf{e}_1 - \mathbf{e}_1 \otimes (\rho_2 - \rho_1) \tag{5.8.3b}$$

$$\mathbf{H} \equiv (\rho_2 - \rho_1) \otimes [(\rho_2 - \rho_1) \times \mathbf{e}_1] \tag{5.8.3c}$$

The interest in developing constraint equations lies in the fact that they are needed to derive systems of independent constrained dynamical equations of mechanical systems. Nonholonomic constraints can also be written in terms of the twists. For instance, let the sphere of the foregoing example be considered as the ith body of a system, the plane being a part of the boundary of the $(i-1)$st body. The arising nonholonomic constraint can then be written as eq.(5.8.1b), with the 3×6 matrices $\mathbf{A}_{i,i-1}$ and $\mathbf{A}_{i,i}$ defined as

$$\mathbf{A}_{i,i-1} \equiv [\,\mathbf{C}_{i-1} \quad \mathbf{1}\,], \quad \mathbf{A}_{i,i} \equiv [\,\mathbf{C}_i \quad \mathbf{1}\,] \tag{5.8.4a}$$

and the Cartesian tensor $\mathbf{C}_i$ is defined as follows:

$$\mathbf{C}_i \equiv \mathbf{1} \times (\mathbf{c}_i - \mathbf{q}) \tag{5.8.4b}$$

with a similar definition for $\mathbf{C}_{i-1}$. In eq.(5.8.4b), $\mathbf{q}$ is the position vector of the contact point between the two bodies.

From eqs.(5.8.2) and (5.8.4) it is apparent that a kinematic constraint, whether holonomic or nonholonomic, can be written in the form of a linear homogeneous equation involving the twists of every pair of coupled bodies of a mechanical system. An essential difference between the holonomic and the nonholonomic constraints previously derived is pointed out here, namely, whereas the former lead to 6-dimensional algebraic equations, the latter lead to 3-dimensional ones. Now, if the mechanical system under study is composed of n coupled rigid bodies, the $6n$ dimensional vector of *generalized twist*, $\mathbf{t}$, is defined as follows:

$$\mathbf{t} \equiv [\, \mathbf{t}_1^T, \mathbf{t}_2^T, \ldots, \mathbf{t}_n^T \,]^T \tag{5.8.5}$$

Furthermore, if the kinematic constraints introduced by the couplings comprise p holonomic and q nonholonomic couplings, then the arising kinematic constraint equations can be described in compact form as follows:

$$\mathbf{A}\mathbf{t} = \mathbf{0} \tag{5.8.6}$$

where $\mathbf{A}$ is an $m \times 6n$ matrix, where $m = 6p + 3q$.

Now the degree of freedom of the system, f, can be determined using the methods of Sections 5.3, 5.4, and 5.7, if the system is composed of bodies coupled by lower kinematic pairs. If not, then the said degree of freedom can be computed as the nullity of $\mathbf{A}$, i.e.,

$$f = \dim(\mathcal{N}) \tag{5.8.7}$$

where $\mathcal{N}$ is the nullspace of $\mathbf{A}$. Hence, f independent variables, henceforth called *independent speeds*, that are grouped in the f-dimensional vector $\mathbf{u}$, can be defined. that produce $\mathbf{t}$ via a linear transformation of the form

$$\mathbf{t} = \mathbf{T}\mathbf{u} \tag{5.8.8}$$

Upon substitution of eq.(5.8.8) into eq.(5.8.6), the following is derived:

$$\mathbf{A}\mathbf{T}\mathbf{u} = \mathbf{0} \tag{5.8.9a}$$

Now, since all components of vector $\mathbf{u}$ are independent functions of time, eq.(5.8.9a) leads to

$$\mathbf{A}\mathbf{T} = \mathbf{0} \tag{5.8.9b}$$

whose right-hand side is the $m \times f$ zero matrix. What eq.(5.8.8b) states is that the columns of $\mathbf{T}$ span the nullspace of $\mathbf{A}$, and hence, $\mathbf{T}$ constitutes an *orthogonal complement* of $\mathbf{A}$.

As shown in Angeles and Lee (1988), matrix $\mathbf{T}$ allows the derivation of f independent Euler-Lagrange dynamical equations of the system under study.

5.9 Analysis of Kinematic Chains

Given an arbitrary f-degree-of-freedom kinematic chain, its analysis consists of determining the motion histories of all its links, given f independent motions of any subset of its links. The foregoing problem has not been solved in its full generality, but a limited class of such analysis problems has been studied extensively. This class of problems is related to simple kinematic chains. As pointed out in Section 5.3, the degree of freedom of a general 7-link closed kinematic chain is 1. The analysis problem associated with this kinematic chain is equivalent to the problem associated with a general 7-link open kinematic chain. Indeed, if the driving link of the closed chain under study is coupled to the fixed link by means of a rotational pair, then the driving link can be regarded as the end link of a similar open chain whose end link undergoes a pure rotation about an axis attached to the fixed link. If the said driving link is coupled to the fixed one via a prismatic pair, then the driving link can be regarded as the end link of an open chain that undergoes a pure translation along the direction of the axis of the said prismatic pair. Thus, it suffices to study the analysis problem of the 7-link open chain, which is conceptually simpler to understand. In fact, the study of the problem associated with the closed chain has been motivated to a great extent by practical problems associated with the open equivalent chain. Freudenstein (1973) is to be cited as the first to have pointed out the theoretical relevance of this problem, whereas Pieper and Roth (Pieper 1968, Pieper and Roth 1969) are among the first researchers to have addressed this problem in a general setting. Prior work had been limited to particular cases in which up to three link lengths vanish, thereby obtaining the popular four-bar linkages. Early work in this area is cited in Pieper (1968). Further work is reported in Roth, Rastegar and Scheinman (1974).

Only as early as 1976 was this problem addressed in its most general form (Albala 1976). In the foregoing reference, Albala reduced the problem under discussion, known as the *inverse kinematic problem* (IKP), to a polynomial equation of the 48th degree in the tangent of half the angle associated with the motion of the driven or output link of a $7R$ closed chain. Duffy and Crane (1980) showed that the IKP leads to a polynomial equation of the 32nd degree in the tangent of half the said angle, whereas Albala showed that his original equation is similarly reduced to a 32nd-degree polynomial equation (Albala 1982). Furthermore, Alizade, Duffy, and Hajiyev (1983-1, 2, 3) and Alizade, Duffy, and Azizov (1983) derived general mathematical models for the analysis and synthesis of four-, five-, six-, and seven-link spatial chains. On the other hand, Tsai and Morgan (1984) conjectured that the IKP should be reducible to a 16th-degree polynomial in the tangent of half the aforementioned angle. More recently,

Primrose (1986) showed that the IKP admits up to 16 solutions, whereas Lee and Liang (1988) derived a polynomial equation of the 16th order whose roots produce the 16 possible different solutions of the IKP. In summary, then, one can say that the IKP admits up to 16 solutions, i.e., for a given pose of the end link, up to 16 different link configurations are possible that produce the same end-link pose. In the discussion which follows, an outline is presented of how the IKP leads to an algebraic problem as discussed previously.

First, the rotation tensor from the $(i-1)$st to the ith link is denoted by $\mathbf{Q}_i$, for $i = 1,\ldots,6$, and the vector directed from a point O_i on the axis of the pair coupling the $(i-1)$st and the ith links to a point O_{i+1} similarly defined, is denoted by $\mathbf{a}_i$. Moreover, the pose of the end link is defined by the position vector $\mathbf{p}$ of a point P of the end link, different from O_n, and tensor $\mathbf{Q}$ defining the orientation of the end link from a certain reference configuration. The kinematic *closure equations* relating tensors $\mathbf{Q}_i$ and vectors $\mathbf{a}_i$ with $\mathbf{Q}$ and $\mathbf{p}$, are the following:

$$\mathbf{Q}_6\mathbf{Q}_5\cdots\mathbf{Q}_1 = \mathbf{Q} \tag{5.9.1a}$$

$$\mathbf{a}_1 + \mathbf{a}_2 + \cdots + \mathbf{a}_6 = \mathbf{p} \tag{5.9.1b}$$

It is pointed out here that, in expressing eq.(5.9.1a) in component form, tensor $\mathbf{Q}_i$ is usually represented in a coordinate frame attached to the ith link, and hence, for reasons that are made apparent in Angeles (1982), the product of the left-hand side of that equation appears in reverse order. In the spirit of the present discussion, however, eq.(5.9.1a) is frame invariant, and hence the order of the tensors appearing in that equation is correct. Furthermore, the notation introduced in Section 5.4 is recalled. Hence, the rotation of link i with respect to link $i-1$ is a function solely of angle θ_i. The same holds for $\mathbf{a}_i$. Thus, eqs.(5.9.1a & b) constitute a nonlinear algebraic system of 12 scalar equations —9 for the first one and 3 for the second one—, out of which the tensor equation contains only 3 independent scalar equations. Indeed, if the left-hand side of that equation is denoted by $\mathbf{P}$, then, due to the orthogonality of the tensors involved, the following holds:

$$\mathbf{P}^T\mathbf{P} = \mathbf{Q}^T\mathbf{Q}$$

which constitutes a system of six independent constraints. The system of eqs.(5.9.1a & b), thus, represents, in fact, six scalar equations in the six unknowns $\theta_1,\ldots,\theta_6$. Since this is a nonlinear system, it may or may not admit a real solution. Moreover, if it admits a real solution, most likely this solution will not be unique. The set of points of the space of joint variables $\{\theta_i\}_1^6$, for which the IKP admits real solutions, is called the *workspace* of the kinematic chain. Of course, since the mapping from joint variables to end-link screw variables—$\mathbf{Q}$ and $\mathbf{p}$—is single valued, the workspace of the chain can also be represented in the space of the latter variables.

From the *implicit-function theorem* (Brand 1955), the workspace of the chain is bounded by the locus of points at which the Jacobian matrix of the set of equations (5.9.1*a* & *b*) becomes rank deficient. In order to derive a system of equations, from which the aforementioned Jacobian matrix can be computed, one can equate the linear invariants of both sides of eq.(5.9.1*a*), thereby deriving four rotation equations which, together with the three translational equations, eq.(5.9.1*b*), produce a system of seven equations of the form:

$$\mathbf{f}(\boldsymbol{\theta}) = \mathbf{s} \tag{5.9.2a}$$

where $\mathbf{f}$ and $\mathbf{s}$ are 7-dimensional vectors and $\boldsymbol{\theta}$ is the 6-dimensional vector whose ith component is θ_i. Moreover, the left-hand side of eq.(5.9.2*a*) contains the unknown $\boldsymbol{\theta}$, whereas the right-hand side, the data, also called the *Cartesian coordinates* of the chain in the specialized literature. Note that the vector of Cartesian coordinates is nothing more than the screw of the end link of the chain, defined as in the second of eqs.(2.5.17). Now, let the Jacobian matrix of $\mathbf{f}$ be denoted by $\mathbf{F}$. Thus,

$$\mathbf{F} = \frac{\partial \mathbf{s}}{\partial \boldsymbol{\theta}} \tag{5.9.2b}$$

On the other hand, from eq.(5.9.2*a*), one can write

$$\mathbf{s} = \mathbf{s}(\boldsymbol{\theta}) \tag{5.9.2c}$$

and hence, by differentiation of both sides of eq.(5.9.2*c*) with respect to time, the following is derived:

$$\dot{\mathbf{s}} = \frac{\partial \mathbf{s}}{\partial \boldsymbol{\theta}}\dot{\boldsymbol{\theta}} \tag{5.9.2d}$$

From eqs.(5.9.2*c* & *d*) it is apparent that the following relation holds:

$$\frac{\partial \mathbf{s}}{\partial \boldsymbol{\theta}} = \frac{\partial \dot{\mathbf{s}}}{\partial \dot{\boldsymbol{\theta}}} \tag{5.9.3}$$

Now, combining eqs.(5.9.2*b*) and (5.9.3), the following is obtained:

$$\mathbf{F} = \frac{\partial \dot{\mathbf{s}}}{\partial \dot{\boldsymbol{\theta}}} \tag{5.9.4}$$

which becomes, upon introduction of the *chain rule*, as follows:

$$\mathbf{F} = \frac{\partial \dot{\mathbf{s}}}{\partial \mathbf{t}}\frac{\partial \mathbf{t}}{\partial \dot{\boldsymbol{\theta}}} \tag{5.9.5}$$

Next, eq.(5.4.7*a*) is recalled. In that equation, $\mathbf{t}_n$ plays the role of $\mathbf{t}$ in the present discussion. Differentiation of both sides of that equation with respect to $\dot{\boldsymbol{\theta}}$ thus yields:

$$\frac{\partial \mathbf{t}}{\partial \dot{\boldsymbol{\theta}}} = \mathbf{J} \tag{5.9.6}$$

On the other hand, from eq.(3.2.23*b*), the following is derived:

$$\frac{\partial \dot{\mathbf{s}}}{\partial \mathbf{t}} = \mathbf{G} \tag{5.9.7a}$$

where $\mathbf{G}$ is the 7×6 matrix defined as

$$\mathbf{G} \equiv \begin{bmatrix} \mathbf{A} & \mathbf{0}_{43} \\ \mathbf{0}_{33} & \mathbf{1} \end{bmatrix} \tag{5.9.7b}$$

Upon substitution of eqs.(5.9.6) and (5.9.7*a*) into eq.(5.9.5), the following is obtained:

$$\mathbf{F} = \mathbf{GJ} \tag{5.9.7c}$$

Hence, apart from the singularities of tensor $\mathbf{1}\mathrm{tr}\mathbf{P} - \mathbf{P}$, which are not intrinsic to the chain, $\mathbf{F}$ becomes singular whenever $\mathbf{J}$ does.

Seven-link chains with neighboring intersecting axes reduce to 6-, 5-, 4- and 3-link chains coupled by lower pairs other than the elementary ones. The IKP associated with many 4-link chains leads to a polynomial equation of the 2nd degree. Some exceptions to these are the *RSRC* and the *RRSC* linkages, whose IKP was studied by Strauchmann and Kassamanian (1977). Yang (1968) reduced the IKP associated with a 5-link *RCRCR* linkage to a 4th-order polynomial equation. Finally, the IKP associated with the so-called wrist-partitioned open chains, equivalent to 5-link *RRRSH* linkages, are known to lead to three cascaded quadratic equations or to a quartic equation cascaded with a quadratic one (Takano 1985).

It is clear that once the joint variables of the closed chain have been determined, the motion of all its linkages is known, thereby completing the analysis problem under study.

5.10 Synthesis of Kinematic Chains

Given the motion. or a family of motions, to be undergone by a subset of links of a kinematic chain, the problem of determining the parameters that define the said chain is known as the *synthesis* problem. Analogous to the analysis problem, this one has not been solved in its full generality. However, a wide class of synthesis problems have been studied and fully solved in the past. For the sake of concreteness, the following discussion will be limited to synthesis problems associated with simple closed kinematic chains. Again, out of these, those with up to three vanishing link lengths, the so-called *four-bar linkages*, have been extensively studied, some solutions being reported for five-, six-, and seven-link chains.

The synthesis problem at hand is known in the specialized literature as the *dimensional-synthesis problem*, to distinguish it from the *type-synthesis and the number-synthesis problems.* The type-synthesis problem refers to the type of mechanical device to be used in order to solve a certain given problem, whereas the number-synthesis problem refers to the number of links to be used to solve the said problem. Neither the type- nor the number-synthesis problems will be discussed here. The dimensional-synthesis problem can be divided into the following types: *i*) function-generation; *ii*) rigid-body guidance; and *iii*) path generation. These are discussed below.

5.10.1 The Function-Generation Synthesis Problem

In this as in the two following problems, the topology of the chain is given, and the dimensions defining the linkage are sought. Thus, a closed chain of a topology given by $K_1 K_2 \cdots K_p$ is specified through its p lower kinematic pairs $\{K_i\}_1^p$, where p can be any number from 2 to 7. Two- and three-link closed chains lead to rather trivial problems and, hence, are not given attention here. Furthermore, the degree of freedom of the chain is assumed to be 1. Now, the linkage parameters to be found are distances and angles between axes of each two consecutive pairs, whereas the purpose of the linkage is to coordinate the motions of the two links coupled to the fixed one. One of these two links is arbitrarily termed the *driving* or *input* link, whereas the other is termed the *driven* or *output* link. Moreover, the input link is assumed to be coupled to the fixed link via a single-degree-of-freedom pair, i.e., either R, P, or H, and hence, its motion is fully defined by one single real variable, which is henceforth denoted by ψ, and can be either an angle or a length supplied with a sign. Note that, if the output link is coupled to the fixed link via a C or an E pair, then either of these can be decomposed into a combination of elementary pairs, R and P, and hence, one single output variable can be defined. Furthermore, S pairs are never used to couple the output link of a function generator with the fixed link. Thus, although the output link may be coupled to the fixed link via a multiple-degree-of-freedom pair, only one real variable, henceforth denoted by ϕ, fully defines the motion of the driven link, by virtue of the assumption that the linkage has a single degree of freedom. If the linkage parameters are now grouped within the n-dimensional vector $\mathbf{p}$, the function-generation synthesis problem can be stated as: *Given a set of m input-output pairs, $\{\psi_i, \phi_i\}_1^m$, determine $\mathbf{p}$ that will define the linkage producing those input-output pairs.*

If, in the foregoing, $m = n$, then the problem may be solved exactly. However, this problem may lead to nonlinear equations, in which case the solution may not be unique. If $m < n$, then there is a deficit of equations and infinitely many solutions may exist. In order to define one of those solutions, an optimization criterion must be imposed. On the contrary, if

$m > n$, then, in general, no solution exists. However, a value of $\mathbf{p}$ can be found that will approximate the given input-output pairs with the minimum error, once this error is defined. The associated synthesis problem for $m = n$ is termed *exact*; it is termed *approximate* for $m > n$. In fact, the exact synthesis problem can be regarded as a particular case of the approximate one, which would yield a zero approximation error.

The most extensively studied class of kinematic chains in this context is that of four-bar linkages. These ones can be further classified into planar, spherical, and spatial. Planar four-bar linkages were studied extensively by Chebyshev in the period comprised between 1861 and 1888 (Artobolevskiĭ et al. 1948), but the formulation of the function-generation synthesis problem associated with this class of linkages was not set into a systematic form suitable for an algebraic formulation until recently (Freudenstein 1955). In fact, Freudenstein showed that, by suitably transforming the link lengths of a planar four-bar linkage—i.e., a planar $RRRR$ closed kinematic chain—into a set of three independent *linkage parameters* k_1, k_2, k_3, which can be grouped within the vector $\mathbf{k}$, then the problem at hand leads to a linear algebraic system of the form

$$\mathbf{A}\mathbf{k} = \mathbf{b} \tag{5.10.1}$$

where $\mathbf{A}$ is an $m \times n$ matrix and $\mathbf{b}$ is an m-dimensional vector, with $n = 3$ for the case under study. Moreover, $\mathbf{A}$ and $\mathbf{b}$ are functions of the given input-output pairs. As a matter of fact, the ith row of $\mathbf{A}$ and the ith component of $\mathbf{b}$ are functions of ψ_i and ϕ_i only. Now, if $m = 3$, then the foregoing algebraic system admits a unique solution given by

$$\mathbf{k} = \mathbf{A}^{-1}\mathbf{b} \tag{5.10.2}$$

provided that $\mathbf{A}$ is nonsingular. Actually, cases rendering $\mathbf{A}$ singular are rather pathological and will normally not appear if the problem is posed properly. If $m > n$, then matrix $\mathbf{A}$ cannot be inverted, properly speaking. However, by defining the approximation error in terms of the Euclidean norm of the error $\mathbf{e}$, defined in turn as $\mathbf{b} - \mathbf{A}\mathbf{k}$, then the value of $\mathbf{k}$ that minimizes the aforementioned norm is given by (Golub and Van Loan 1983):

$$\mathbf{k} = (\mathbf{A}^T\mathbf{A})^{-1}\mathbf{A}^T\mathbf{b} \tag{5.10.3}$$

Hartenberg and Denavit (1964) extended Freudenstein's idea to spherical and spatial linkages. For spherical linkages, they showed that a mapping of the link dimensions—angles in this case—into four linkage parameters $k_1, \cdots, k_4$ leads to a linear algebraic system of the form of eq.(5.10.1), with $n = 4$. Hence, what was said about planar linkages can be extended to spherical ones, with obvious modifications. Four-bar spatial linkages are more varied but, in general, suitable transformations of their dimensions—distances and angles—into a set of linkage parameters $k_1, \cdots, k_p$ can be found that will lead to a linear algebraic system of the form of eq.(5.10.1). For instance, for the $RSSR$ linkage, p is known to be 6 (Hartenberg and Denavit 1964), whereas for the $RCCC$ linkage, $p = 8$ if the output is the linear motion of the driven link; $p = 5$ if the said output is the angular motion of the same link.

In defining the approximation error of the synthesis equations for the problem at hand, the actual error between the prescribed value of the output variable, ϕ_i, and the synthesized value, say $\overline{\phi}_i$, has not been measured directly, but rather through the synthesis equations. In other words, if the foregoing error, $\phi_i - \overline{\phi}_i$, is denoted by ϵ_i, then $\mathbf{e}$ is, in fact, a nonlinear vector function of the set $\{\epsilon_i\}_1^m$, grouped within the m-dimensional vector ϵ. Whereas ϵ_i is known as the *structural error* at the ith configuration, e_i, the ith component of $\mathbf{e}$, has been termed the *design error* (Tinubu and Gupta 1984) at the same configuration. If the ith input-output equation is regarded as a nonlinear one in variable ϕ_i, of the form:

$$f_i(\psi_i, \phi_i; \mathbf{k}) = 0 \tag{5.10.4a}$$

then the m synthesis equations can be written in vector form as

$$\mathbf{f}(\mathbf{u}, \mathbf{v}; \mathbf{k}) \equiv \mathbf{b} - \mathbf{A}\mathbf{k} = \mathbf{0} \tag{5.10.4b}$$

in which $\mathbf{u}$ and $\mathbf{v}$ are m-dimensional vectors whose ith components are ψ_i and ϕ_i, respectively. Now, $\mathbf{f}$ is immediately identified in eq.(5.10.4b) as the design error $\mathbf{e}$. Hence, the structural error and the design error are related via eq.(5.10.4b) in the form

$$\mathbf{e} = \mathbf{e}(\epsilon) \equiv \mathbf{b} - \mathbf{A}\mathbf{k} \tag{5.10.4c}$$

where $\mathbf{e}$ is a nonlinear function of ϵ. Hence, it is now evident that minimizing a norm of the design error is not equivalent to minimizing that norm of the structural error, and vice versa. Thus, if the structural error is to be minimized, its gradient with respect to the vector of linkage parameters, $\mathbf{k}$, is required. This is computed via the gradient of the design error with respect to $\mathbf{k}$, by recalling *the chain rule*, i.e.,

$$\frac{d\epsilon}{d\mathbf{k}} = \frac{d\epsilon}{d\mathbf{e}}\frac{d\mathbf{e}}{d\mathbf{k}} \tag{5.10.5a}$$

where, from the definition of $\mathbf{e}$,

$$\frac{d\mathbf{e}}{d\mathbf{k}} = -\mathbf{A} \tag{5.10.5b}$$

and $d\epsilon/d\mathbf{e}$ is derived as follows: Function $\mathbf{f}$ defined in eq.(5.10.4b) can be regarded as

$$\mathbf{f} \equiv \mathbf{f}(\mathbf{e}, \epsilon) = \mathbf{0} \tag{5.10.6a}$$

which defines implicitly ϵ as a function of $\mathbf{e}$. Next, differentiation of $\mathbf{f}$ with respect to $\mathbf{e}$ leads to

$$\frac{d\mathbf{f}}{d\mathbf{e}} = \frac{\partial \mathbf{f}}{\partial \mathbf{e}} + \frac{\partial \mathbf{f}}{\partial \epsilon}\frac{d\epsilon}{d\mathbf{e}} = \mathbf{0} \tag{5.10.6b}$$

in which all matrices involved are of $m \times m$. If $\partial \mathbf{f}/\partial \epsilon$ is nonsingular, then eq.(5.10.6*b*) can be solved for $d\epsilon/d\mathbf{e}$, thereby obtaining:

$$\frac{\partial \epsilon}{\partial \mathbf{e}} = -\left(\frac{\partial \mathbf{f}}{\partial \epsilon}\right)^{-1}\frac{\partial \mathbf{f}}{\partial \mathbf{e}} \tag{5.10.6c}$$

in which

$$\frac{\partial \mathbf{f}}{\partial \mathbf{e}} = \mathbf{1}, \quad \frac{\partial \mathbf{f}}{\partial \epsilon} = \frac{\partial \mathbf{f}}{\partial \mathbf{v}} \tag{5.10.6d}$$

where, in accordance with the definition given before, $\mathbf{v}$ is the following array:

$$\mathbf{v} \equiv [\, \psi_1, \ldots, \psi_m \,]^T$$

and hence, eq.(5.10.6*c*) reduces to

$$\frac{d\epsilon}{d\mathbf{e}} = -(\frac{\partial \mathbf{f}}{\partial \mathbf{v}})^{-1} \tag{5.10.6e}$$

Substitution of eqs.(5.10.5*b*) and (5.10.6*e*) into eq.(5.10.5*a*) leads to

$$\frac{d\epsilon}{d\mathbf{k}} = (\frac{\partial \mathbf{f}}{\partial \mathbf{v}})^{-1}\mathbf{A} \tag{5.10.7}$$

which is the relation sought. In the foregoing discussion, it has been assumed that both the input and the output variables are measured from preassigned references. However, if these are left undefined, then they have to be determined from the synthesis equations, which now become nonlinear for the said reference values appear in matrix $\mathbf{A}$—and in vector $\mathbf{b}$ as well. Luck (1976) first showed that the two aforementioned reference values obey a certain relation, and gave a method to solve for one in terms of the other—and of the given input-output pairs—, for the exact-synthesis problem. Later, Angeles (1986-3) extended the method to the approximate-synthesis problem. As a matter of fact, the introduction of the said reference values allows for the synthesis of a family of linkages producing the prescribed input-output pairs with the least-square error. That is, for every real value that one of the two reference values can attain, one such linkage can be defined, which allows for the choice of the best in some sense, while imposing certain performance characteristics on the linkage performance. In this regard, the goodness of the linkage can be measured in terms other than the approximating error. One item that arises naturally in this context is the force-transmission characteristic, which measures the amount of applied generalized force that is used to produce useful work—as opposed to that resulting in nonworking constraint forces. A generally applicable *transmission index*, first proposed by Sutherland and Roth (1973), amounts to the absolute value of the cosine of the angle between the 6-dimensional vector of the twist of the output link and of the wrench applied on it by its moving neighbor—note that the output link has two neighbors, namely, one that is fixed and another that is moving. Based on this, a *transmission quality* was defined for any four-bar linkage in Angeles and Bernier (1987-2).

On the other hand for a wide class of four-bar linkages, the input-output equation is quadratic (Angeles and Bernier 1987-1), i.e., it takes on the following form:

$$A(\psi)T^2 + 2B(\psi)T + C(\psi) = 0 \tag{5.10.8a}$$

in which A, B, and C contain the linkage parameters, grouped in vector $\mathbf{k}$, as parameters, and $T \equiv \tan(\phi/2)$, where both the input and the output variables are assumed to be angular. In some instances, the said input-output equation can be quartic (Strauchmann and Kassamanian 1977). In this case, the input-output equation takes on the form:

$$A(\psi)T^4 + B(\psi)T^3 + C(\psi)T^2 + D(\psi)T + E(\psi) = 0 \tag{5.10.8b}$$

whose coefficients $A, \cdots, E$ are functions of the linkage parameters, as well as of the input variable ψ. If, for a given vector $\mathbf{k}$ of linkage parameters, the input-output equation, either (5.10.5) or (5.10.6), admits real roots, for any value of ψ in the interval $[\,0,\ 2\pi\,]$, then the input link is said to possess *full rotatability*, and is called a *crank*. If the said input-output equation admits real roots only for a limited subinterval $I_\psi \in [\,0,\ 2\pi\,]$, then the input link has partial rotatability, and is called a *rocker*. Moreover, if the input-output equation does not admit any real solution for any value of ψ, then the linkage is not feasible, i.e., it cannot be assembled into a closed kinematic chain. Determining whether the input or the output link of a linkage is capable of full rotatability is a problem known as *mobility analysis*. This problem, as pertaining to planar four-bar linkages, was already addressed by Grashof (1883), who derived mobility criteria for the full rotatability of any of the four links of a planar four-bar linkage. Later, Duditza and Dittrich (1969) derived mobility criteria applicable to spherical four-bar linkages, similar to those of Grashof's, whereas Geise and Modler (1981) derived similar criteria for spatial four-bar linkages. Mobility criteria for general four-bar linkages whose input-output equation can be reduced to a quadratic equation are given in Angeles and Bernier (1987-1), whereas Williams II and Reinholtz (1987) outline a method for deriving mobility criteria for spatial four-bar linkages whose input-output equation is quartic.

Now, the transmission quality Q of a simple, closed, single-degree-of-freedom kinematic chain is defined in terms of its transmission index, $I(\psi)$.

This is first derived in terms of the twist of its output link, $\mathbf{t}_P$, and the wrench applied to it by the moving link to which it is coupled, $\mathbf{w}_P$. These are defined as

$$\mathbf{t}_P \equiv \begin{bmatrix} \omega \\ \mathbf{v}_P \end{bmatrix}, \quad \mathbf{w}_P \equiv \begin{bmatrix} \mathbf{n}_P \\ \mathbf{f} \end{bmatrix} \tag{5.10.9a}$$

where ω and $\mathbf{v}_P$ denote the angular velocity of the output link and the velocity of a point P of the axis of the pair coupling the output link with its moving neighbor. Moreover, $\mathbf{n}_P$ and $\mathbf{f}$ denote the moment about and the force applied at P that the said neighbor exerts on the output link. Then, $I(\psi)$ is defined as

$$I(\psi) \equiv \left| \frac{\mathbf{w}_P^T \mathbf{W} \mathbf{t}_P}{\|\mathbf{V}\mathbf{w}_P\| \, \|\mathbf{V}\mathbf{t}_P\|} \right| \tag{5.10.9b}$$

where $\mathbf{W}$ is a 6×6 positive definite weight matrix which is introduced for purposes of rendering the inner product of the numerator of the right-hand side expression of eq.(5.10.9b) dimensionally homogeneous. Furthermore, since $\mathbf{W}$ is positive definite, it admits the following factoring:

$$\mathbf{W} = \mathbf{V}^T \mathbf{V} \tag{5.10.9c}$$

The 6×6 matrix $\mathbf{V}$ appearing in the denominator of the right-hand side expression of eq.(5.10.9b) is thus the factor of $\mathbf{W}$ appearing in eq.(5.10.9c).

Clearly, $I(\psi)$ is configuration dependent, i.e., it is a function of ψ, and attains values comprised between 0 and 1. However, a measure of the global transmission characteristic of the linkage should be configuration independent. Such a measure, then, can be defined as a norm of $I(\psi)$. If the Euclidean norm, now associated with a Hilbert space, is used to define such a measure, and the input link is assumed to move so that ψ attains values between ψ_a and $\psi_a + \Delta\psi$, then the aforementioned norm will be termed the *transmission quality* of the linkage, which will be denoted by Q, and is defined as

$$Q \equiv \sqrt{\frac{1}{\Delta\psi} \int_{\psi_a}^{\psi_a+\Delta\psi} I^2(\psi) d\psi} \tag{5.10.9d}$$

Clearly, Q is bounded by 0 and 1 but, by virtue of the continuity of the input-output equations of simple closed kinematic chains, it never attains these values, i.e., 0 and 1 are, in fact, the *greatest lower bound* and the *lowest upper bound* of Q, for any linkage.

Finally, the synthesis problem for function generation using a 5-, 6-, or 7-link closed kinematic chain, as well as that associated with complex chains, has been solved only for some particular cases (Nitescu 1982; Felzien and Cronin 1985; Sandgren 1985; Sandor, Kohli and Zhuang 1985). A major problem with this regard is that, whereas four-bar linkages with quadratic input-output equations admit two different configurations for each value of the input variable ψ, called the *conjugate configurations* of the linkage, other 4-bar linkages, as well as some 5-bar linkages, admit four conjugate configurations. The general 7-link closed chain admits up to 16 conjugate configurations. Hence, this multiplicity of solutions poses a serious algebraic difficulty when solving these problems, and a general methodology applicable to the solution of these is still a subject under research.

5.10.2 Kinematic Synthesis for Rigid-Body Guidance

Again, for the sake of concreteness, only simple closed kinematic chains will be considered. In this problem, a discrete set of poses $\{Q_i\}_0^m$ of a link is specified via the associated screws $\{\mathbf{s}_i\}_1^m$, where $\mathbf{s}_i$ describes the motion of the said link from the 0th pose Q_0 to Q_i, for $i = 1, \ldots, m$. Moreover, this link is assumed to be other than those coupled to the fixed ones, which is why it is usually referred to as a *floating link*. In 4-bar linkages, there is one single floating link, which is then called the *coupler link*. As in the preceding problem, the 4-link kinematic chain has been studied in great depth, but some results on 5-, 6-, and 7-link chains can be found in the literature. The problem under study can be formulated as follows: *Given a discrete set of m poses of a particular floating link of a simple closed kinematic chain of a given topology, determine the dimensions of the chain that has a floating link attaining the prescribed poses.* Burmester (1886) is to be cited as the researcher who first addressed this problem. He did so in connection with the synthesis of the planar four-bar linkage. The methodology meant to solve the associated problem is known as *the Burmester Theory*. Within the framework of the Burmester Theory, the coupler link to be guided through the specified poses is assumed to be coupled to a link which is in turn coupled to the fixed link, all couplings taking place via rotational pairs. The link coupled to the fixed and the coupler links is termed a *dyad*, which is defined by its two end points, A and B. These points are in fact the centers of the pairs coupling the dyad to the fixed and the coupler links, respectively. Hence, A is said to be a fixed joint center, whereas B is a moving joint center. Now, let B_i denote the position of B at the ith pose of the coupler link, for $i = 0, 1, \ldots, m$, whereas $\mathbf{a}$ and $\mathbf{b}_i$ denote the 2-dimensional position vectors of A and B_i. Moreover, let each pose be defined via a point R_i, of position vector $\mathbf{r}_i$, and an angle θ_i, for $i = 0, 1, \ldots, m$. If, in particular, θ_0 is arbitrarily defined as 0, then θ_i, for $i = 1, 2, \ldots, m$ is the angle of rotation from the 0th pose to the ith pose. It can be shown—see, for instance, (Roth 1967-1)—that $\mathbf{b}_i$ can be written as follows:

$$\mathbf{b}_i = \mathbf{r}_i + \mathbf{Q}_i(\mathbf{b}_0 - \mathbf{r}_0) \tag{5.10.10}$$

where $\mathbf{Q}_i$ is the 2-dimensional rotation from the 0th pose to the ith one.

Since the dyad is a rigid body, the distance between B_i and A remains constant throughout the motion, and hence, one can write:

$$\|\mathbf{b}_i - \mathbf{a}\|^2 = \|\mathbf{b}_0 - \mathbf{a}\|^2, \quad i = 1, \ldots, m \tag{5.10.11a}$$

or, if eq.(5.10.10) is taken into account,

$$\|\mathbf{r}_i + \mathbf{Q}_i(\mathbf{b}_0 - \mathbf{r}_0) - \mathbf{a}\|^2 = \|\mathbf{b}_0 - \mathbf{a}\|^2, \quad i = 1, \ldots, m \tag{5.10.11b}$$

After performing some simplifications, eq.(5.10.11b) reduces to

$$2(\mathbf{r}_i - \mathbf{a})^T \mathbf{Q}_i(\mathbf{b}_0 - \mathbf{r}_0) + 2\mathbf{a}^T \mathbf{b}_0 - 2\mathbf{r}_i^T \mathbf{a} - 2\mathbf{r}_0^T \mathbf{b}_0 + \|\mathbf{r}_0\|^2 = 0,$$
$$\text{for } i = 1, \ldots, m \tag{5.10.12}$$

Equation (5.10.12) represents a system of m scalar equations in four unknowns, namely, the two components of $\mathbf{a}$ and the two components of $\mathbf{b}_0$. Moreover, each of these equations is nonlinear—in fact, quadratic—in the unknowns. Thus, if $m = 4$, the problem may be solved exactly, but the arising solution may not be unique. Next, an outline of the solution procedure to determine both vectors $\mathbf{a}$ and $\mathbf{b}_0$ is given. It is first noticed that eq.(5.10.12) is, in fact, *bilinear* in both $\mathbf{a}$ and $\mathbf{b}_0$. Hence, two of the four foregoing equations can be solved for either of these. For instance, $\mathbf{b}_0$ can be solved for in terms of $\mathbf{a}$. This would yield each component of $\mathbf{b}_0$ as a rational function of the coordinates of $\mathbf{a}$, whose numerator is a linear polynomial in the components of $\mathbf{a}$, its denominator being a quadratic polynomial in the same variables. This result is due to the fact that each entry of the matrix coefficient of the arising linear algebraic system is linear in $\mathbf{a}$, which renders the determinant of the said 2×2 matrix a quadratic polynomial in $\mathbf{a}$. Furthermore, substitution of the arising expression for $\mathbf{b}_0$ in the other two equations would yield two cubic equations in the components of $\mathbf{a}$, i.e., in the coordinates of point A. These two cubic equations can be reduced to one single ninth-degree equation in one of the said coordinates. However, as shown in Bottema and Roth (1979), each cubic equation can be reduced to a quadratic, and hence, the solution of the two equations leads, in fact, to a quartic equation. Thus, the problem admits either zero, two, or four real solutions. If two such solutions exist, then each of these constitutes a dyad, and the ensemble of the two dyads, each coupled to the fixed and the coupler links, thus defines one single linkage. If four solutions exist, then the four arising dyads can be combined in six different pairs, thereby obtaining up to six possible linkages whose coupler link attains the five specified poses–the 0th one included. Whereas a multiplicity of solutions may exist for this problem, it is pointed out that the arising linkages may contain a coupler link that goes through a subset of the five specified poses in one conjugate configurations and through the remaining poses in the other one. That is, it may happen that the synthesized linkage, although having joints located at points whose position vectors verify exactly the four synthesis equations, cannot go through all five specified poses, unless it is disassembled and reassembled in its conjugate set of configurations.

This problem has been extensively studied for $m < 4$, i.e., for less than five prescribed poses. For $m = 3$, or four specified poses, one of the four involved coordinates should be prescribed. Next, one of the three arising synthesis equations can be solved for the other coordinate of the same point. For instance, if the x-coordinate of B_0 is prescribed as a given real number, say b, then its y-coordinate can be determined from one of those equations, in terms of the coordinates of A, as shown below. The first of the synthesis equations, together with the condition that the x-coordinate of B_0 be b, can be written as the following linear algebraic system of two equations in the two components of vector $\mathbf{b}_0$:

$$\begin{bmatrix} 2[\mathbf{a} - \mathbf{r}_0 + \mathbf{Q}_1^T(\mathbf{r}_1 - \mathbf{a})]^T \\ \mathbf{i}^T \end{bmatrix} \mathbf{b}_0 = \begin{bmatrix} 2[(\mathbf{r}_1 - \mathbf{a})^T \mathbf{Q}_1 \mathbf{r}_0 + \mathbf{r}_1^T \mathbf{a}] - \|\mathbf{r}_0\|^2 \\ b \end{bmatrix} \tag{5.10.13}$$

From eq.(5.10.13) it is apparent that the two components of $\mathbf{b}_0$ are rational functions of two polynomials of first degree in $\mathbf{a}$–the x-component is, in fact, a very simple rational function, namely, b itself. Substitution of the arising expression for $\mathbf{b}_0$ into the other two synthesis equations—for $i = 2, 3$—produces two quadratic equations in $\mathbf{a}$, from which the two coordinates of A can be computed. Clearly, as b takes on real values from $-\infty$ to $+\infty$, the coordinates of both A and B_0 vary continuously between $-\infty$ and $+\infty$. The locus of all possible positions of A is called the *center-point curve*, whereas that of all possible positions of B_0 is called the *circle-point curve*. The reason for these names lies in the fact that, as the linkage moves, B_0 describes a circular trajectory centered at A.

If $m = 2$, i.e., if three poses of the coupler link are specified, then the two coordinates of one of the two points sought, either A or B_0, have to be prescribed. For instance, if $\mathbf{a}$ is prescribed, then the two synthesis equations are linear in $\mathbf{b}_0$ and the solution of these for this unknown is straightforward and unique.

The Burmester Theory was extended to the synthesis of spatial dyads by Roth and collaborators (Roth 1967-1; Roth 1967-2; Chen and Roth 1969-1; Chen and Roth 1969-2; Tsai and Roth 1972, Tsai and Roth 1973). In this case, many types of dyads arise, i.e., as many as the possible combinations of the six different lower kinematic pairs taking two at a time, with the provision that a dyad, say RC, is to be distinguished from CR. A comprehensive study of the design of these dyads appears in Tsai and Roth (1972). Since RR dyads bear a particular significance, for reasons that will soon become apparent, these were devoted special attention in Tsai and

Roth (1973). An alternative approach to the problem, that resembles that of planar dyads previously discussed, is given in Angeles (1982), and is at this point briefly recalled. Again, one of the two rotational pairs of the dyad will be considered fixed and the other one moving. Furthermore, let the axis of the fixed pair be parallel to the unit vector $\mathbf{u}$, that of the moving dyad being parallel to the unit vector $\mathbf{v}_0$ when the conducted rigid body is in its 0th pose. At the ith pose, the moving axis is assumed to be parallel to the unit vector $\mathbf{v}_i$. Additionally, similar to the planar case, a point A of the fixed axis and a point B of the moving one totally define the location of these axes. The position vector of A is denoted by $\mathbf{a}$, whereas that of B in the ith pose by $\mathbf{b}_i$. As shown in Angeles (1982), the rigidity condition of the dyad leads to exactly the same equation, eq.(5.10.12), as that derived for the planar dyad, except that now all vectors are 3-dimensional and the rotation tensor is of 3×3. Thus, eq.(5.10.12) is, again, the synthesis equation for the problem at hand. However, additional equations have to be introduced in order to account for the fact that the problem is 3-dimensional. Moreover, as compared to the planar problem, the problem at hand has additional unknowns. Indeed, vectors $\mathbf{u}$ and $\mathbf{v}_0$ are to be determined, in addition to $\mathbf{a}$ and $\mathbf{b}_0$, which increases the number of unknowns to twelve, i.e., three for each of the four foregoing vectors. Additional equations are next derived. First, vectors $\mathbf{u}$ and $\mathbf{v}_0$ are of unit magnitude, and hence,

$$\mathbf{u}^T\mathbf{u} = 1 \tag{5.10.14a}$$

$$\mathbf{v}_0^T\mathbf{v}_0 = 1 \tag{5.10.14b}$$

Furthermore, points A and B on the pair axes are assumed to be the intersections of the common perpendicular to those axes with them, and hence,

$$\mathbf{u}^T(\mathbf{b}_i - \mathbf{a}) = 0, \quad i = 0, 1, \ldots, m \tag{5.10.15a}$$

$$\mathbf{v}_i^T(\mathbf{b}_i - \mathbf{a}_0) = 0, \quad i = 0, 1, \ldots, m \tag{5.10.15b}$$

But, of course,

$$\mathbf{v}_i = \mathbf{Q}_i\mathbf{v}_0, \quad i = 1, \ldots, m \tag{5.10.16}$$

Upon substitution of eqs.(5.10.10) and (5.10.16) into eq.(5.10.15b), the following is obtained:

$$\mathbf{v}_0^T\mathbf{Q}_i^T[\mathbf{Q}_i(\mathbf{b}_0 - \mathbf{r}_0) + \mathbf{r}_i] = 0, \quad i = 0, 1, \ldots, m \tag{5.10.17}$$

However, the perpendicularity conditions (5.10.15a) and (5.10.17) do not suffice to guarantee that the rigidity of the dyad is observed. In fact, these equations could hold but $\mathbf{v}_i$ could, at the same time, reverse its sign, which would amount to a reflection of the dyad at the ith pose, rather than a rotation. In order to ensure a rotation from the 0th to the ith pose, then, the following equations are to be observed:

$$\mathbf{u}^T\mathbf{v}_i = \mathbf{u}^T\mathbf{v}_0, \quad i = 1, \ldots, m \tag{5.10.18}$$

In summary, then, the synthesis equations are eqs.(5.10.12), (5.10.14a & b), (5.10.15a), (5.10.17), and (5.10.18), i.e., $4m + 4$ equations to solve for the previously described twelve unknowns. Thus, if the number of equations is to equal that of unknowns, then $m = 2$, which amounts to a nonlinear algebraic system of twelve equations in twelve unknowns. Therefore, it is possible, in principle, to design an RR dyad that will guide a rigid body exactly through three specified poses and, moreover, the problem being nonlinear, may admit multiple solutions. Roth (1967-2) showed that this problem cannot admit more than twenty-four solutions, whereas Suh (1969) showed that these solutions always come in pairs that constitute an $RRRR$ spatial linkage of the Bennett type (Bennett 1903). Later, Tsai and Roth (1973) showed that this problem leads to a cubic equation and that it admits exactly two real roots, which therefore constitute a Bennett mechanism.

An alternate approach to the foregoing method of dyad synthesis, regarding the problem under study, is to synthesize the overall kinematic chain. This requires, then, the introduction of the closure equations (5.9.1a & b) and others pertaining to the types of kinematic pairs that are being used. This approach has been successfully applied to the synthesis of planar four-bar linkages for approximate rigid-body guidance, i.e., with $m > 4$ (Akhras and Angeles 1987).

Furthermore, Hunt (1978) presents a detailed study of the Burmester Theory as applied to spherical linkages. Finally, Sandor, Xu, and Weng (1987) report the synthesis of a $7R$ closed kinematic chain for rigid-body guidance, whereas Sandor, Kohli, and Zhuang (1985) reported the solution to a problem associated with a complex spatial kinematic chain for rigid-body guidance. In the problem of $7R$ synthesis, branching becomes a major concern, for up to sixteen conjugate branches may appear.

5.10.3 Kinematic Synthesis for Path Generation

This problem is similar to the previous one, except that, in this case, the orientation of the guided body is not specified, the variables defining the said orientation thus becoming unknowns in the problem. Again, for the sake of brevity, only four-bar linkages will be considered in connection with this problem. In this regard, the problem consists of determining the dimensions of a linkage of a given topology which contains a coupler link with one point that describes a path passing through a set of given positions. This path is known as a *coupler curve*. Coupler curves of planar four-revolute linkages were studied comprehensively by Hartenberg and Denavit (1964), whereas a study of coupler curves of spherical four-revolute linkages was reported by Dittrich and Zakel (1975). As to spatial four-bar linkages, Suh and Radcliffe (1978) present a method of solution that they apply to an $RRSS$ four-bar linkage, and allows them to find a linkage with a coupler curve passing through a set of eight prescribed points. Furthermore, the same authors applied their method to the synthesis of an $RRSS - SS$ linkage—i.e., a linkage with two closed kinematic loops, one of which is an $RRSS$ linkage, the other being otained by appending and SS dyad to the coupler link of the former—, thereby allowing for the synthesis of a linkage whose coupler link has a point tracing a path that passes through up to eleven points.

Again, this synthesis problem has been studied most extensively in connection with planar four-bar linkages. In this regard, for $m+1$ specified locations of a point R of the coupler link, the orientation of this in the 0th pose is irrelevant and the angle θ_0 defining it can be assigned any value, e.g., 0. Therefore, all successive orientations of the coupler link are given with respect to the 0th pose. As to the equations available, they are the same as those derived for the synthesis of a planar RR dyad, except that now the equations for the two dyads constituting the kinematic chain are coupled, and hence, two sets of equations (5.10.12) have to be written, one for each dyad, which leads to $2m$ equations. Now, the number of unknowns is $8+m$, namely, the 8 coordinates of the four points defining the two dyads plus the m values θ_i, for $i = 1, \ldots, m$. If the number of equations is to equal that of the unknowns, then, $m = 8$, which means that an exact synthesis problem of this type can be solved by specifying nine locations of point R of the coupler. However, Jensen (1984) showed that, if considerations of symmetry are exploited in synthesizing symmetric coupler curves, then up to twelve locations of point R can be met exactly. If, in general, more than eight locations are specified, the problem becomes one of approximate synthesis. The problem of approximate synthesis for path generation has received extensive attention in connection with planar four-bar linkages (Sutherland and Karwa 1978; Klein 1981; Paradis and Willmert 1983).

As an alternative approach, independent of the dyad-synthesis method, the problem has been solved by writing the closure equations of the overall closed chain. In fact, this approach lends itself better to the approximate synthesis problem, and is the one that has been applied more extensively (Sutherland and Karwa 1978; Klein 1981; Paradis and Willmert 1983). As in the case of rigid-body guidance, branching defect may appear when solving this problem, whereby every synthesized linkage verifying the synthesis equations should be tested for this defect. Associated defects that arise in synthesizing linkages for any of the three synthesis problems are the *order defect* and the *Grashof defect*. The first defect arises whenever the synthesized linkage has a coupler with a point R that, although it passes through the prescribed positions either with zero or with a minimum error, it does so in an order different from the prescribed one. The second defect arises whenever the synthesized linkage turns out to have an input link of the rocker type, although one of the crank type is needed. Waldron and Stevensen, Jr. (1979) have proposed some solutions to overcome the foregoing defects when dealing with planar four-bar linkages.

Finally, Ravani and Roth (1983, 1984) introduced the concept of kinematic mapping that allows to formulate problems of rigid-body guidance as those of curve fitting, thereby leading to a unified formulation of both rigid-body guidance and path generation.

In summary, an outline of problems in kinematic synthesis has been introduced for the sake of completeness. However, solution methods in connection with these problems which are still a subject of intensive research have not been discussed, for these fall outside of the scope of this book. Nevertheless, a few references have been included for the interested reader who wishes to pursue a more detailed study of the subject.

REFERENCES

Akhras, R. and Angeles, J.(1987) "Unconstrained nonlinear least-square optimization of planar linkages for rigid-body guidance", Department of Mechanical Engineering and Robotic Mechanical Systems Laboratory-McRCIM Technical Report, McGill University, Montreal.

Albala, H.(1976) *Displacement Analysis of the N-Bar, Single-Loop, Spatial Linkage. Application to the 7R Single-Degree-of-Freedom Spatial Mechanism*, Doctor of Science Thesis, Technion-Israel Institute of Science and Technology, Haifa.

Albala, H.(1982) "Displacement analysis of the general n-bar, single-loop, spatial linkage. Part I: Underlying mathematics and useful tables. Part II: Basic displacement equations in matrix and algebraic form", *Trans. ASME Journal of Mechanical Design* (104) 2: 504–525.

Alizade, R. I., Duffy, J. and Azizov, A. A. (1983) "Mathematical models for analysis and synthesis of spatial mechanisms–II. Five-link spatial mechanisms", *Mechanism and Machine Theory* (18) 5: 309–315.

Alizade, R. I., Duffy, J. and Hajiyev, E. T. (1983-1) "Mathematical models for analysis and synthesis of spatial mechanisms–I. Four-link spatial mechanisms", *Mechanism and Machine Theory* (18) 5: 301–307.

Alizade, R. I., Duffy, J. and Hajiyev, E. T. (1983-2) "Mathematical models for analysis and synthesis of spatial mechanisms–III. Six-link spatial mechanisms", *Mechanism and Machine Theory* (18) 5: 317–322.

Alizade, R. I., Duffy, J. and Hajiyev, E. T. (1983-3) "Mathematical models for the analysis and synthesis of spatial mechanisms–IV. Seven-link spatial mechanisms", *Mechanism and Machine Theory* (18) 5: 323–328.

Angeles, J.(1982) *Spatial Kinematic Chains. Analysis, Synthesis, Optimization*, Springer-Verlag, Berlin-Heidelberg-New York.

Angeles, J.(1985) "On the numerical solution of the inverse kinematic problem", *The International Journal of Robotics Research*, (4) 2: 21–37.

Angeles, J.(1986-1) "Automatic computation of the screw parameters of rigid-body motions. Part I: Finitely-separated positions", *Trans. ASME Journal of Dynamic Systems, Measurement, and Control* (108) 6: 32–38.

Angeles, J.(1986-2) "Automatic computation of the screw parameters of rigid-body motions. Part II: Infinitesimally-separated positions", *Trans. ASME Journal of Dynamic Systems, Measurement, and Control* (108) 6: 39–43.

Angeles, J.(1986-3) "Optimierung ebener, sphärischer und räumlicher Getriebe zur approximierten Lagenzuordnung", *Mechanism and Machine Theory* (21) 2: 187–197.

Angeles, J. (1987) "Computation of rigid-body angular acceleration from point-acceleration measurements", *Trans. ASME Journal of Dynamic Systems, Measurement, and Control* (109) 2: 124–127.

Angeles, J. (1988) "Special loci of the workspace of spherical wrists", *Proc. International Meeting on Robot Kinematics*, Sept. 19–21, Ljubljana, YU.

Angeles, J. and Gosselin (1988) "Détermination du degré de liberté des chaînes cinématiques simples et complexes", to appear in *Transactions of the Canadian Society of Mechanical Engineering.*

Angeles, J. and Bernier, A.(1987-1) "A general method of four-bar linkage mobility analysis", *Trans. ASME Journal of Mechanisms, Transmissions, and Automation in Design* (109) 2: 197–203.

Angeles, J. and Bernier, A.(1987-2) "The global least-square optimization of function-generating linkages", *Trans. ASME Journal of Mechanisms, Transmissions, and Automation in Design* (109) 2:204–209.

Angeles, J. and Lee, S.(1988) "The formulation of dynamical equations of holonomic mechanical systems using a natural orthogonal complement", *ASME Journal of Applied Mechanics* (55) 1: 243–244.

Angeles, J. and López-Cajún, C.(1988) "The dexterity index of serial-type robotic manipulators", *Proc. 20th Biennial Mechanisms Conference*, Sept. 25–28, Kissimmee, FL: 79–84.

Angeles, J. and Rojas, A.(1987) "Manipulator inverse kinematics via condition-number minimization and continuation", *The International Journal of Robotics and Automation* (2) 2: 61–69.

Artobolevskiĭ, I. I.(1975) *Teoriîa Mekhanizmov i Mashin* (Theory of Mechanisms and Machines), Nauka Publishers, Moscow.

Artobolevskiĭ, I. I. et al.(1948) *Polnoe Sobranie Sochineniĭ P. L. Chebysheva. Tom IV: Teoriîa Mekhanizmov*, (Complete Collected Works of P. L. Chebyshev. Vol. IV: Theory of Mechanisms), Akademiîa Nauk SSSR Publisher, Moscow-Leningrad.

Baker, J. E.(1986) "Limiting positions of a Bricard linkage and their possible relevance in the cyclohexane molecule", *Mechanism and Machine Theory* (21) 3: 253–260.

Ball, Sir R. S.(1900) *Theory of Screws*, Cambridge University Press, Cambridge.

Bennett, G. T.(1903) "A new mechanism", *Engineering*, Vol. 76: 777–778.

Bottema, O. and Roth, B.(1979) *Theoretical Kinematics*, North-Holland Publishing Co., Amsterdam.

Bowen, R. M., and Wang, C.-C.(1976) *Introduction to Vectors and Tensors. Vol. I*, Plenum Press, New York.

Brand, L.(1955) *Advanced Calculus*, John Wiley and Sons, Inc., New York-London-Sydney.

Bricard, R.(1927) *Leçons de Cinématique*, Gauthier-Villars, Paris.

Burmester, L.(1886) *Lehrbuch der Kinematik*, Verlag von Arthur Felix, Leipzig.

Chen, P. and Roth, B.(1969-1) "A unified theory for the finitely and infinitesimally separated position problems of kinematic synthesis", *Trans. ASME Journal of Engineering for Industry* (91) 1: 203–208.

Chen, P. and Roth, B.(1969-2) " Design equations for the finitely and infinitesimally separated position synthesis of binary links and combined link chains", *Trans. ASME Journal of Engineering for Industry* (91) 1: 209–219.

Chiang, C. H.(1988) *Kinematics of Spherical Mechanisms*, Cambridge University Press, Cambridge-New York-New Rochelle-Melbourne-Sydney.

Coddington, E. A., and Levinson, N.(1955) *Theory of Ordinary Differential Equations*, McGraw-Hill Book Co., New York.

Coriolis, G.(1835) "Mémoire sur les équations du mouvement relatif des systèmes des corps", *J. Ecole Polytechnique* (15) 24: 142–154.

Dijksman, E. A.(1976) *Motion Geometry of Mechanisms*, Cambridge University Press, Cambridge.

Dimentberg, F. M.(1978) *Teoriĭa Vintov i ee Prilozheniĭa* (The Theory of Screws and Its Applications), Nauka, Moscow.

Dittrich, G. and Braune, R.(1978) *Getriebetechnik in Beispiele* , Oldenburg Verlag, Munich.

Dittrich, G. and Zackel, H.(1975) "Sphärische Koppelkurven und ihre Anwendungen", *Proc. 4th World Congress on Theory of machines and Mechanisms*, Newcastle-upon-Tyne (4):939–944.

Duditza, Fl. and Dittrich, G.(1969) "Die Bedingungen für die Umlauffähigkeit sphärischer viergliedriger Kurbelgetriebe", *Industrie-Anzeiger* (91) 71: 1687–1690.

Duffy, J.(1980) *Analysis of Mechanisms and Robot Manipulators*, Edward Arnold, London.

Duffy, J. and Crane, C.(1980) "A displacement analysis of the general spatial 7-link, 7R mechanism", *Mechanism and Machine Theory*, (15) 3: 153–159.

Erdman, A. G. and Sandor, G. N.(1984) *Mechanism Design: Analysis and Synthesis*, Vol. 1, Prentice-Hall, Inc., Englewood Cliffs.

Ericksen, J. L.(1960) "Tensor Fields", in S. Flügge (ed.), *Encyclopedia of Physics*, Vol. III/1, Springer-Verlag, Berlin-Göttingen-Heidelberg.

Euler, L.(1750) "Recherches sur l'effet d'une machine hydraulique par Mr. Segner Professeur à Göttingue", *Mémoires de l'Académie des Sciences de Berlin* (6): 311–354 = *Opera Omnia* (2) 15: 1–39.

Euler, L.(1758) "Du mouvement de rotation des corps solides autour d'un axe variable", *Mémoires de l'Académie des Sciences de Berlin* (14): 154–193 = *Opera Omnia* (2) 8: 200–235.

Euler, L.(1765) "Theoria motus corporum solidorum seu rigidorum ex primis nostrae cognitionis principiis stabilita et ad omnes motus, qui in huiusmodi corpora cadere possunt, accomodata", Rostock: 3–293 = *Opera Omnia* (2) 3: 3–327 and 4: 3–358.

Euler, L.(1776), "Nova methodus motum corporum rigidorum determinandi", *Novii Comentarii Academiæ Scientiarum Petropolitanæ*, 20 (1775) 1776: 208–238 = *Opera Omnia* (2) 9: 99–125.

Felzien, M. L. and Cronin, D. L.(1985) "Steering error optimization of the McPherson strut automotive suspension", *Mechanism and Machine Theory* (20) 1: 17–26.

Finkbeiner, II, D. T.(1966) *Introduction to Matrices and Linear Transformations*, 2nd. ed., W. H. Freeman and Company, San Francisco and London.

Freudenstein, F.(1955) "Approximate synthesis of four-bar linkages", *Trans. ASME* (77): 853–861.

Freudenstein, F.(1973) "Kinematics: Past, Present and Future", *Mechanism and Machine Theory* (8) 2: 151–160.

Geise, G. and Modler, K.(1981) "Umlauffähigkeit räumlicher Mechanismen", *Mechanism and Machine Theory* (16) 6: 653–659.

Golub, G. H. and van Loan, C.(1983) *Matrix Computations*, The Johns Hopkins University Press, Baltimore.

Gosselin, C.(1988) *Kinematic Analysis, Optimization and Programming of Parallel Robotic Manipulators*, Ph. D. Thesis, McGill University, Montreal.

Grashof, F.(1883) *Theoretische Maschinenlehre*, Vol. 2, Verlag L. Voss, Leipzig.

Halmos, P. R.,(1974) *Finite-Dimensional Vector Spaces*, Springer-Verlag, New York.

Hamel, G.(1912) *Elementare Mechanik*, Leipzig and Berlin.

Hamilton, W. R.(1899) *Elements of Quaternions*, Cambridge University Press, Cambridge.

Harary, F.(1972) *Graph Theory*, Addison-Wesley Publishing Company, Reading, MA.

Hartenberg, R. S., and Denavit, J.(1964) *Kinematic Synthesis of Linkages*, McGraw-Hill Book Co., New York.

Hennenberg, L.(1903) "Die graphische Statik der starren Körper", in Klein, F. and Müller, C. (eds.), *Encyclopädie der Mathematischen Wissenschaften* Vol. IV, 1, Druck und Verlag von B. G. Teubner, Leipzig: 349–434.

Hervé, J. M.(1978) "Analyse structurelle des mécanismes par groupes de déplacements", *Mechanism and Machine Theory*, Vol. 13: 437–450.

Ho, C. Y.(1978) "A note on the existence of Bennett mechanism", *Mechanism and Machine Theory* (13) 3: 269–271.

Hunt, K. H.(1978) *Kinematic Geometry of Mechanisms*, Clarendon Press, Oxford.

Ishlinskiĭ, A. Yu.(ed.)(1967) *Razvitie Mekhaniki v SSSR* (The Development of Mechanics in the USSR), Nauka Publishers, Moscow.

Jamalov, R. I., Litvin, F. L. and Roth, B.(1984) "Analysis and design of *RCCC* linkages", *Mechanism and Machine Theory* (19) 4/5: 397–407.

Jensen, P. W.(1984) "Synthesis of four-bar linkages with a coupler point passing through 12 points", *Mechanism and Machine Theory* (19) 1: 149–156.

Jung, G.(1908) "Geometrie der Massen", in Klein, F. and Müller, C. (editors) *Encyclopädie der Mathematischen Wissenschaften*, Vol. IV, 1, Druck un Verlag von B. G. Teubner, Leipzig: 282–344.

Kane, T. R.(1961) "Dynamics of nonholonomic systems", *Trans. ASME Journal of Applied Mechanics* (83): 574–578.

Kane, T. R.(1968) *Dynamics*, Holt, Rinehart and Winston, New York.

Kane, T. R., Likins P. W., and Levinson D. A.(1983) *Spacecraft Dynamics*, McGraw-Hill Book Co., New York.

Karger, A. and Novák, J.(1978) *Prostorová Kinematika a Lieovy Grupy* (in Czech), SNTL, Prague. In English: (1985) *Space Kinematics and Lie Groups*, Gordon and Breach Science Publishers, New York-London-Paris-Montreux.

Kelland, P., and Tait, P. Q.(1882) *Introduction to Quaternions*, MacMillan and Co., London.

Kirchhoff, G.(1876) *Vorlesungen über Mathematische Physik: Mechanik*, Leipzig.

Klein B.(1981) "Zum Einsatz nichtlinearer Optimierungsverfahren zur rechnerunterstützten Konstruktion ebener Koppelgetriebe", *Mechanism and Machine Theory* (16) 5: 567–576.

Konstantinov, M. S., Vrigazov, A. G., Stanchev, E. S., and Nedelchev, I. N. (1980) *Teoriĭâ na Mekhanizmite i Mashinite*, Drjavno Publishers, Sophia.

Lee, H.-Y. and Liang, C.-C.(1988) "Displacement analysis of the general spatial 7-link 7*R* mechanism", submitted to *Mechanism and Machine Theory*.

Lichtenheldt, W., and Luck, K.(1979) *Konstruktionslehre der Getriebe*, Akademie-Verlag, Berlin.

Luck, K.(1976) "Computersynthese des viergliedrigen räumlichen Koppelgetriebes vom Typ RSSR", *Mechanism and Machine Theory* (11) 3: 213–225.

Meyer zur Capellen, W.(1941) *Mathematische Instrumente*, Akademische Verlagsanstalt, Becker und Erler, K. G., Leipzig.

Myard F. E.(1931) "Contribution à la géometrie des systèmes articulés", *Bulletin de la Société Mathématique de France* (59): 183–210.

Nieto, J. (1978) *Síntesis de Mecanismos*, Editorial AC, Madrid.

Nikravesh, P. E., Wehage, R. A. and Kwon, O. K.(1985) "Euler parameters in computational kinematics and dynamics. Part 1", *Trans. ASME J. Mechanisms, Transmissions, and Automation in Design* (107): 358–365.

Nitescu, P. N.(1982) "On the kinematic synthesis of 5-axis spatial mechanisms", *Mechanism and Machine Theory* (17) 6: 387–395.

Paradis, M. J. and Willmert, K. D.(1983) " Optimal mechanism design using the Gauss constrained method", *Trans. ASME Journal of Mechanisms, Transmissions, and Automation in Design* (105) 2: 187–196.

Pauli W.(1958) "Die allgemeinen Prinzipien der Wellenmechanik", in Flügge S.(editor) *Encyclopedia of Physics*, Vol. V/1, Springer-Verlag, Berlin-Göttingen-Heidelberg.

Pelecudi, Chr.(1967) *Bazele Analizei Mecanismelor*, Ed. Acad. R. S. R., Bucharest.

Phillips, J.(1984) *Freedom in Machinery, Vol. 1. Introducing Screw Theory*, Cambridge University Press, Cambridge-London- New York-New Rochelle-Melbourne-Sydney.

Pieper, D. L.(1968) *The Kinematics of Manipulators under Computer Control*, Ph. D. Thesis, Stanford University, Stanford, CA.

Pieper, D. L. and Roth, B.(1969) "The kinematics of manipulators under computer control", *Proc. 2nd International Congress on Theory of Machine and Mechanisms* (2): 159–169.

Primrose, E. J. F.(1986) "On the input-output equation of the general 7*R* mechanism", *Mechanism and Machine Theory* (21) 6: 509–510.

Ravani, B. and Roth, B.(1983) "Motion synthesis using kinematic mappings", *Trans. ASME Journal of Mechanisms, Transmissions, and Automation in Design* (105) 3: 460–467.

Ravani, B. and Roth, B.(1984) "Mappings of spatial kinematics", *Trans. ASME Journal of Mechanisms, Transmissions, and Automation in Design* (106) 3: 341–347.

Reuleaux, F.(1875) *Theoretische Kinematik, Grundzüge einer Theorie des Maschinenwesens*, 1. Teil, Verlag Friedrich Vieweg & Sohn, Braunschweig.

Rodrigues, O.(1840) "Des lois géométriques qui régissent les déplacements d'un solide dans l'éspace, et de la variation des coordonnées provenant de ces déplacements considérés independamment des causes qui peuvent les produire", *Journal de Mathématiques Pures et Appliquées* (5) 1st. Series: 380–440.

Roth, B.(1967-1) "The kinematics of motion through finitely separated positions", *Trans. ASME Journal of Applied Mechanics* (89) 3: 591–598.

Roth, B.(1967-2) "Finite-position theory applied to mechanism synthesis", *Trans. ASME Journal of Applied Mechanics* (89): 3 599–605.

Roth, B., Rastegar, J. and Scheinman, V.(1974) "On the design of computer controlled manipulators", *On Theory and Practice of Robots and Manipulators* (1) Springer-Verlag, New York: 93–113.

Salisbury, J. K. and Craig, J. J.(1982) "Articulated hands: force control and kinematic issues", *The International Journal of Robotics Research* (1) 1:4–17.

Sandgren, E.(1985) "Design of single- and multiple-dwell six-link mechanisms through design optimization", *Mechanism and Machine Theory* (20) 6: 483–490.

Sandor, G. N. and Erdman, A. G.(1984) *Advanced Mechanism Design: Analysis and Synthesis*, Vol. 2, Prentice-Hall, Englewood Cliffs.

Sandor, G. N., Kohli, D. and Zhuang, X.(1985) "Synthesis of *RSSR-SRR* spatial motion generator mechanism with prescribed crank rotations for three and four finite positions", *Mechanism and Machine Theory* (20) 6: 503–519.

Sandor, G. N., Xu, Y. and Weng, T.-C.(1987) "Synthesis of 7-*R* spatial motion generators with prescribed crank rotations and elimination of branching", *The International Journal of Robotics Research* (5) 2: 143–156.

Schoenflies, A. and Grübler, M.(1902) *Kinematik*, in Klein, F. and Müller, C. (eds.), *Encyklopädie der Mathematischen Wissenschaften*, Vol. IV, 1, Druck und Verlag von B. G. Teubner, Leipzig: 192–278.

Schuman, H.(1952) *Leonardo da Vinci on the Human Body: The Anatomical, Physiological, and Embryological Drawings of Leonardo da Vinci*, Henry Schuman, New York.

Spring, K.(1986) "Euler parameters and the use of quaternion algebra in the manipulation of finite rotations: A review", *Mechanism and Machine Theory* (21) 5: 365–373.

Stäckel, P.(1908) "Elementare Dynamik der Punktsysteme und Starren Körper", in Klein, F. and Müller, C. (eds.) *Encyklopädie der Mathematischen Wissenschaften*, Vol. IV, 1, Druck und Verlag von B. G. Teubner, Leipzig: 443–684.

Stewart, D.(1965) "A platform with six degrees of freedom", *Proc. Institution of Mechanical Engineers* (180) 5: 371–378.

Strauchmann, H. and Kassamanian, A. A.(1977) "Rechnergestützte Analyse des *RSRC*-Mechanismus", *Wissenschaftliche Zeitschrift der Technischen Universität Dresden* (26) 5: 899–903.

Study, E.(1903) *Die Geometrie der Dynamen*, Leipzig.

Suh, C. H.(1969) "On the duality in the existence of R-R links for three positions", *Trans. ASME Journal of Engineering for Industry* (91) 1: 129–134.

Sutherland, G. H. and Karwa, N. R.(1978) "Ten-design-parameter 4-bar synthesis with tolerance considerations", *Mechanism and Machine Theory* (13) 3: 311–327.

Sutherland, G. H. and Roth, B.(1973) "A transmission index for spatial mechanisms", *Trans. ASME Journal of Engineering for Industry* (95) 2: 589–597.

Szabó, I.(1977) *Geschichte der mechanischen Prinzipien*, Birkhäuser Verlag, Basel and Stuttgart.

Takano, M.(1985) "A new effective solution to inverse kinematics problem (synthesis) of a robot with any type of configuration", *Journal of the Faculty of Engineering*, The University of Tokyo (B) (38) 2: 107–135.

Timerding H. E.(1902) "Geometrische Grundlegung der Mechanik eines starren Körpers" in Klein, F. and Müller, C. (eds.), *Encyklopädie der Mathematischen Wissenschaften*, Vol. IV, 1, Druck und Verlag von B. G. Teubner, Leipzig: 128–189.

Tinubu, S. O. and Gupta, K. C.(1984) "Optimal synthesis of function generators without the branch defect", *Trans. ASME Journal of Mechanisms, Transmissions, and Automation in Design* (106) 3: 348–354.

Truesdell, C.(1966) *The Elements of Continuum Mechanics*, Springer-Verlag, New York.

Truesdell, C. and Toupin, R. A.(1960) "The Classical Field Theories", in Flügge, S. (ed.), *Encyclopedia of Physics*, Vol. III/1, Springer-Verlag, Berlin-Göttingen-Heidelberg.

Tsai, L. W. and Morgan, A. P.(1984) "Solving the kinematics of the most general six- and five-degree-of-freedom manipulators by continuation methods", *Trans. ASME Journal of Mechanisms, Transmissions, and Automation in Design* (107) 2: 189–200.

Tsai, L. W. and Roth, B.(1972) " Design of dyads with helical, cylindrical, spherical, revolute and prismatic joints", *Mechanism and Machine Theory* (7) 1: 85–102.

Tsai, L. W. and Roth, B.(1973) "A note on the design of revolute-revolute cranks", *Mechanism and Machine Theory* (8) 1: 23–31.

Veldkamp, G. R. (1969) "Acceleration axes and distribution in spatial motion", *Trans. ASME Journal of Engineering for Industry* (91) 1: 147–151.

Voinea, R. P. and Atanasiou, M. C. (1962) "Théorie géométrique des vis et quelques applications à la théorie des mécanismes", *Revue de Mécanique Appliquée* (7): 845–860.

Volmer, J. (ed.) (1979) *Getriebetechnik-Lehrbuch*, 4th Edition, VEB Verlag Technik, Berlin.

Waldron, K. J. and Stevensen, Jr., E. N.(1979) "Elimination of branch, Grashof, and order defects in path-angle generation and function generation synthesis", *Trans. ASME Journal of Mechanical Design* (101) 3: 428–437.

Wang, C.-C.(1979) *Mathematical Principles of Mechanics and Electromagnetism. Part A: Analytical and Continuum Mechanics*, Plenum Press, New York and London.

Whittaker, E. T.(1927) *A Treatise on the Analytical Dynamics of Particles and Rigid Bodies*, Cambridge University Press, Cambridge.

Williams II, R. L. and Reinholtz, C. F.(1987) "Mechanism link rotatability and limit position analysis using polynomial discriminants", *Trans. ASME Journal of Mechanisms, Transmissions, and Automation in Design* (109) 2: 178–182.

Wittenburg J.(1977) *Dynamics of Systems of Rigid Bodies*, B. G. Teubner, Stuttgart.

Woo, L. and Freudenstein, F. (1970) "Applications of line geometry to theoretical kinematics and the kinematic analysis of mechanical systems", *Journal of Mechanisms* (5) 3: 417–460.

Yang, A. T.(1968) "Displacement analysis of spatial five link mechanisms using (3×3) matrices with dual-number elements", *Trans. ASME Journal of Engineering for Industry* (91) 1: 152–164.

INDEX